Advanced Materials and Technologies in Endodontics and Restorative Dentistry

Advanced Materials and Technologies in Endodontics and Restorative Dentistry

Guest Editors

Luigi Generali
Vittorio Checchi
Eugenio Pedullà

Basel • Beijing • Wuhan • Barcelona • Belgrade • Novi Sad • Cluj • Manchester

Guest Editors

Luigi Generali	Vittorio Checchi	Eugenio Pedullà
Department of Surgery, Medicine, Dentistry and Morphological Sciences With Transplant Surgery, Oncology and Regenerative Medicine Relevance	Department of Surgery, Medicine, Dentistry and Morphological Sciences With Transplant Surgery, Oncology and Regenerative Medicine Relevance	Department of General Surgery and Medical Surgical Specialties University of Catania Catania Italy
University of Modena & Reggio Emilia	University of Modena & Reggio Emilia	
Modena	Modena	
Italy	Italy	

Editorial Office
MDPI AG
Grosspeteranlage 5
4052 Basel, Switzerland

This is a reprint of the Special Issue, published open access by the journal *Materials* (ISSN 1996-1944), freely accessible at: www.mdpi.com/journal/materials/special_issues/H644Q8G8D1.

For citation purposes, cite each article independently as indicated on the article page online and using the guide below:

Lastname, A.A.; Lastname, B.B. Article Title. *Journal Name* **Year**, *Volume Number*, Page Range.

ISBN 978-3-7258-3730-4 (Hbk)
ISBN 978-3-7258-3729-8 (PDF)
https://doi.org/10.3390/books978-3-7258-3729-8

Cover image courtesy of Luigi Generali
SEM-FEG micrographs showing presence of bacteria inside dentinal tubules (12000 x).

© 2025 by the authors. Articles in this book are Open Access and distributed under the Creative Commons Attribution (CC BY) license. The book as a whole is distributed by MDPI under the terms and conditions of the Creative Commons Attribution-NonCommercial-NoDerivs (CC BY-NC-ND) license (https://creativecommons.org/licenses/by-nc-nd/4.0/).

Contents

Eugenio Pedullà, Francesco Saverio Canova, Giusy Rita Maria La Rosa, Alfred Naaman, Franck Diemer and Luigi Generali et al.
Influence of NiTi Wire Diameter on Cyclic and Torsional Fatigue Resistance of Different Heat-Treated Endodontic Instruments
Reprinted from: *Materials* 2022, 15, 6568, https://doi.org/10.3390/ma15196568 1

Emad Youssef, Holger Jungbluth, Søren Jepsen, Manfred Gruener and Christoph Bourauel
Comparing Cyclic Fatigue Resistance and Free Recovery Transformation Temperature of NiTi Endodontic Single-File Systems Using a Novel Testing Setup
Reprinted from: *Materials* 2024, 17, 566, https://doi.org/10.3390/ma17030566 10

Hayate Unno, Arata Ebihara, Keiko Hirano, Yuka Kasuga, Satoshi Omori and Taro Nakatsukasa et al.
Mechanical Properties and Root Canal Shaping Ability of a Nickel–Titanium Rotary System for Minimally Invasive Endodontic Treatment: A Comparative In Vitro Study
Reprinted from: *Materials* 2022, 15, 7929, https://doi.org/10.3390/ma15227929 23

Joanna Falkowska, Tomasz Chady, Włodzimierz Dura, Agnieszka Droździk, Małgorzata Tomasik and Ewa Marek et al.
The Washout Resistance of Bioactive Root-End Filling Materials
Reprinted from: *Materials* 2023, 16, 5757, https://doi.org/10.3390/ma16175757 35

Rubén Herrera-Trinidad, Pedro Molinero-Mourelle, Manrique Fonseca, Adrian Roman Weber, Vicente Vera and María Luz Mena et al.
Assessment of pH Value and Release of Calcium Ions in Calcium Silicate Cements: An In Vitro Comparative Study
Reprinted from: *Materials* 2023, 16, 6213, https://doi.org/10.3390/ma16186213 51

Yun-Jae Ha, Donghee Lee and Sin-Young Kim
The Combined Effects on Human Dental Pulp Stem Cells of Fast-Set or Premixed Hydraulic Calcium Silicate Cements and Secretome Regarding Biocompatibility and Osteogenic Differentiation
Reprinted from: *Materials* 2024, 17, 305, https://doi.org/10.3390/ma17020305 67

Seong-Hee Moon, Seong-Jin Shin, Seunghan Oh and Ji-Myung Bae
Antibacterial Activity and Sustained Effectiveness of Calcium Silicate-Based Cement as a Root-End Filling Material against *Enterococcus faecalis*
Reprinted from: *Materials* 2023, 16, 6124, https://doi.org/10.3390/ma16186124 83

Arne Peter Jevnikar, Tine Malgaj, Kristian Radan, Ipeknaz Özden, Monika Kušter and Andraž Kocjan
Rheological Properties and Setting Kinetics of Bioceramic Hydraulic Cements: ProRoot MTA versus RS+
Reprinted from: *Materials* 2023, 16, 3174, https://doi.org/10.3390/ma16083174 94

Tiago Reis, Cláudia Barbosa, Margarida Franco, Ruben Silva, Nuno Alves and Pablo Castelo-Baz et al.
Three-Dimensional Printed Teeth in Endodontics: A New Protocol for Microcomputed Tomography Studies
Reprinted from: *Materials* 2024, 17, 1899, https://doi.org/10.3390/ma17081899 108

Xiaoming Zhu, Jiamin Shi, Xinyi Ye, Xinrong Ma, Miao Zheng and Yang Yang et al.
Influence of Cold Atmospheric Plasma on Surface Characteristics and Bond Strength of a Resin Nanoceramic
Reprinted from: *Materials* **2022**, *16*, 44, https://doi.org/10.3390/ma16010044 . . . . **123**

Article

Influence of NiTi Wire Diameter on Cyclic and Torsional Fatigue Resistance of Different Heat-Treated Endodontic Instruments

Eugenio Pedullà [1], Francesco Saverio Canova [1], Giusy Rita Maria La Rosa [1,*], Alfred Naaman [2], Franck Diemer [3,4], Luigi Generali [5,†] and Walid Nehme [2,†]

[1] Department of General Surgery and Medical-Surgical Specialties, University of Catania, Via Santa Sofia, 78, 95123 Catania, Italy
[2] Endodontic Department, Saint Joseph University of Beirut, Rue de Damas P.O. Box 17-5208, Beirut 1104 2020, Lebanon
[3] Restorative and Endodontic Department, CHU de Toulouse, University of Toulouse, 31013 Toulouse, France
[4] Clement Ader Institute, UMR CNRS 5312, 31400 Toulouse, France
[5] Endodontic Section, Department of Surgery, Medicine, Dentistry and Morphological Sciences with Transplant Surgery, Oncology and Regenerative Medicine Relevance (CHIMOMO), School of Dentistry, University of Modena and Reggio Emilia, 41125 Modena, Italy
* Correspondence: g_larosa92@live.it
† These authors contributed equally to this work.

Abstract: We compared the mechanical properties of 2Shape mini TS2 (Micro-Mega, Besançon, France) obtained from 1.0 diameter nickel-titanium (NiTi) wires and 2Shape TS2 from 1.2 diameter nickel-titanium (NiTi) wires differently thermally treated at room and body temperature. We used 120 NiTi TS2 1.0 and TS2 1.2 files made from controlled memory (CM) wire and T-wire (n = 10). Cyclic fatigue resistance was tested by recording the number of cycles to fracture (NCF) at room and body temperatures using a customized testing device. Maximum torque and angle of rotation at failure were recorded, according to ISO 3630-1. Data were analyzed by a two-way ANOVA (p < 0.05). The CM-wire files had significantly higher NCFs at both temperatures, independent of wire dimensions. Testing at body temperature negatively affected cyclic fatigue of all files. The 1.0-mm diameter T-wire instruments showed higher NCF than the 1.2-mm diameter, whereas no significant differences emerged between the two CM wires at either temperature. The maximum torque was not significantly different across files. The TS2 CM-wire files showed significantly higher angular rotation to fracture than T-wire files. The TS2 CM-wire prototypes showed higher cyclic fatigue resistance than T-wire prototypes, regardless of wire size, exhibiting suitable torsional properties. Torsional behavior appears to not be affected by NiTi wire size.

Keywords: cyclic fatigue; endodontics; heat treatment; NiTi wire diameter; torsional fatigue; 2Shape

1. Introduction

Since endodontic instruments have been manufactured with nickel-titanium [NiTi] wire, the use of NiTi files has largely increased in endodontics [1]. NiTi endodontic instruments possess extreme flexibility and strength, even if they are vulnerable to fracture during use in clinical situations [2]. The main causes leading to fracture of NiTi files are torsional and cyclic fatigue [3]. Fracture caused by torsional fatigue occurs when the file engages the root canals as the handpiece continues to rotate. Torsional failure is defined by a maximum torsional load and angle of rotation. This last property is connected to the file's ability to twist before fracture [2]. Cyclic fatigue occurs when the file rotates freely within the canal, and is exposed to repeated cycles of compression and traction in the part of the root canal with the greatest curvature [3]. Although both failure modes simultaneously

Citation: Pedullà, E.; Canova, F.S.; La Rosa, G.R.M.; Naaman, A.; Diemer, F.; Generali, L.; Nehme, W. Influence of NiTi Wire Diameter on Cyclic and Torsional Fatigue Resistance of Different Heat-Treated Endodontic Instruments. *Materials* **2022**, *15*, 6568. https://doi.org/10.3390/ ma15196568

Academic Editor: Edgar Schäfer

Received: 30 August 2022
Accepted: 19 September 2022
Published: 22 September 2022

Publisher's Note: MDPI stays neutral with regard to jurisdictional claims in published maps and institutional affiliations.

Copyright: © 2022 by the authors. Licensee MDPI, Basel, Switzerland. This article is an open access article distributed under the terms and conditions of the Creative Commons Attribution (CC BY) license (https:// creativecommons.org/licenses/by/ 4.0/).

occur in clinical situations, many studies have found that cyclic fatigue fracture is the principal cause of file separation [4].

Several factors, including file size, cross-sectional area, design, heat treatment, and metallurgical properties of the instruments, affect the mechanical properties of rotary files [5]. Recent studies have also shown that temperature is another factor influencing the mechanical behavior of NiTi instruments. A significant decrease in the cyclic fatigue strength of the instruments was found when the files were tested at body temperature [6–8].

In this context, heat treatments have been proposed to prevent the fracture of NiTi files and improve the metallurgical properties of these instruments, modifying the austenite finish temperature [9]. Above this temperature, instruments exist completely as austenitic structures that results in more rigidity than martensitic ones [10,11]. Some examples of heat treatment technology include controlled memory wire (CM; Coltene, Cuyahoga Falls, OH, USA) [12] and T-wire (Micro-Mega, Besançon, France). CM-wire heat treatment is a distinctive process that modifies the memory of the material [13], whereas T-wire treatment, according to the manufacturer, improves flexibility and cyclic fatigue resistance [14]. One of the file systems constituted from T-wire alloy is 2Shape (TS) (Micro-Mega) [1]. The TS file system possesses a triple-helix cross-section and is produced in two sizes: TS1 (size #25/0.04) and TS2 (size #25/0.06) [5]. In addition, TS2 files can be produced from a 1.0-mm or 1.2-mm diameter NiTi wire, which generates TS2 mini (1.0) and TS2 (1.2) files, respectively. Both TS2 mini 1.0 and TS2 1.2 files have the same tip and taper (25.06) but differ in the amount of metal due to the different diameters of the NiTi wire used. To date, no studies have compared the cyclic fatigue resistance of NiTi TS2 mini 1.0 and TS2 1.2 rotary files or T-wire and CM-wire 2Shape TS2 files, because the latter ones are available as prototypes.

Therefore, this study aims to evaluate and compare the cyclic and torsional fatigue resistance of differently heat-treated (i.e., T-wire and CM-wire) TS2 mini 1.0 and TS2 1.2 files. The null hypotheses were: (i) the wire diameter does not affect fatigue resistance of tested instruments at room and body temperature; (ii) the wire diameter does not interfere in the maximum torque load or angular rotation of instruments with different heat treatments.

2. Materials and Methods

No human or animal subjects were used in this study; therefore, ethics committee approval was not required [15]. Based on the results of similar previous studies [2,4], sample size estimation was calculated a priori with G*Power 3.1.9.2 software (Heinrich-Heine- University at Dusseldorf, Dusseldorf, Germany) to have 80% power and alpha error probability of 0.05.

The 2Shape TS2 mini 1.0-mm and TS2 1.2-mm files, differently heat-treated (i.e., T-wire and prototypes CM-wire), were used. All files were 25 mm long, with 10 instruments in each group used in cyclic fatigue and torsional resistance tests. Prior to testing, all files were inspected using a stereomicroscope (SZR- 10; Optika, Bergamo, Italy) to verify any deformation [2]. None were discarded.

2.1. Cyclic Fatigue Test

Twenty instruments from each of these systems were tested in vitro with a static cyclic fatigue test at room ($25 \pm 1\ °C$) and body ($37 \pm 1\ °C$) temperatures, for a total of 80 NiTi rotary instruments. Cyclic fatigue testing was performed using a custom machine (Figure 1) with a stationary unit that maintained a 6:1 electric reduction handpiece (Sirona Dental Systems GmbH, Bensheim, Germania) in a fixed three-dimensional position and a movable rail mount that allowed file positioning within the artificial canal. All instruments were placed in a precise and reproducible manner using the electric handpiece [5] (Figure 2).

The mobile platform contained a 16 mm-long stainless steel artificial canal, with an angle of 60 degrees and a radius of 5 mm curvature [15]. The artificial canal was specifically designed for the TS2 instrument employed in terms of size (25) and taper (0.06), giving it a sufficient lumen with suitable trajectory. For a better clinical simulation [16,17], the temperature adjustment was guaranteed by a thermostat which ensured the proper heating of the artificial canal containing the instrument. The temperature was maintained constantly during the test by means of a thermocouple applied to the artificial canal, which activated or deactivated the thermostatic resistance when the temperature decreased or reached the preset one, respectively. The TS2 mini 1.0 and TS2 1.2 were tested using a continuous rotation at 300 rpm (revolutions per minute) and at the maximum torque (4.1 Ncm) until they broke [5] (Figure 3). A special high-flow synthetic oil (Super Oil; Singer Co Ltd., Elizabeth, NJ) was applied as a lubricant. The time to fracture was recorded with a stopwatch and confirmed by video, recorded using a digital camera [18]. The number of cycles to fracture (NCF) was also calculated using the equation "rpm/60 × time to fracture (seconds)" [18].

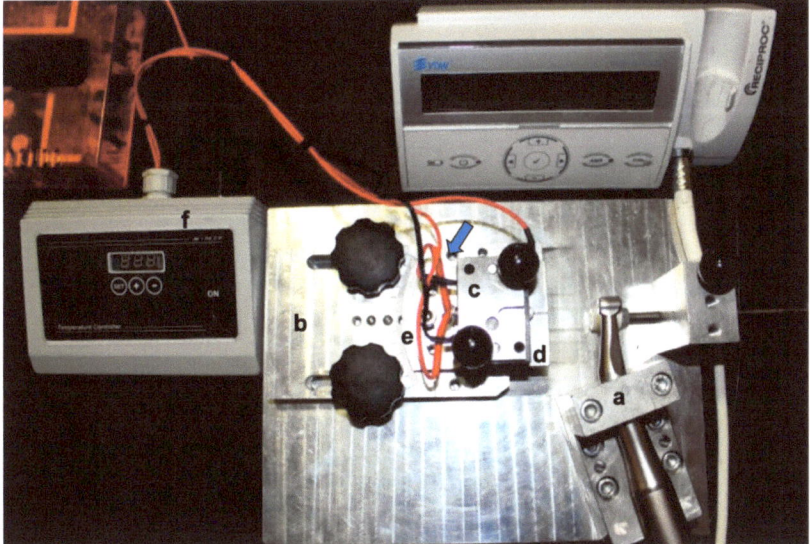

Figure 1. An illustrative figure of the customized testing device employed for cyclic fatigue tests. (a) The electric handpiece was maintained in a stable position by a block system; (b) a mobile support on rails allowed the insertion/withdrawal of the NiTi file in a standardized manner in the 0.06 tapered artificial canal (c); (d) a mobile platform permitted to put the file in different inclinations marked by the angles reported on the mobile support (all files were tested at 0°) (e); (f) a thermostat was used to check the temperature with an acceptable variation of ±1 °C. The blue arrow indicates the thermocouple applied to the artificial canal.

2.2. Torsional Fatigue Test

For the torsional test, each instrument was clamped at 3 mm from the tip by a chuck connected to a torque-sensing load cell; after this, the shaft of the file was fastened into an opposing chuck, able to be rotated with a stepper motor in the clockwise direction at a speed of 2 revolutions per minute until file breakage. The torque load (Ncm) and angular rotation (°) were monitored continuously by a torsiometer (Sabri Dental Enterprises, Downers Grove, IL, USA) at room temperature (21 °C ± 1 °C), and the ultimate torsional strength and angle of rotation at failure were recorded.

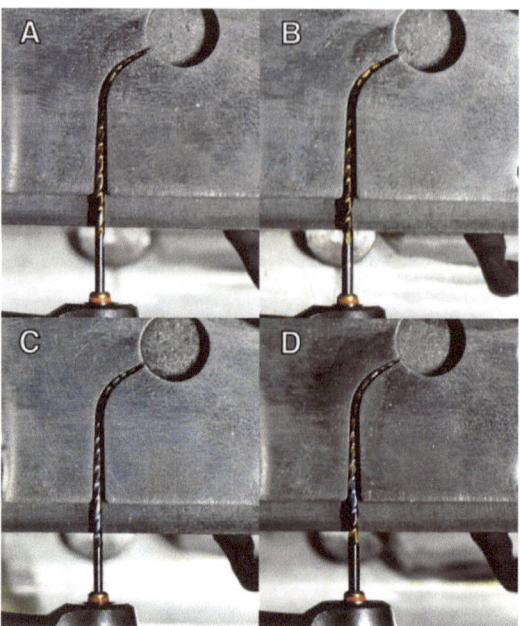

Figure 2. (**A**) TS2 T-wire 1.0, (**B**) 1.2, and (**C**) CM-wire 1.0, (**D**) 1.2 in the artificial canal with 60° angle of curvature and 5-mm radius at body temperature.

Figure 3. Part of instrument using TS2 CM-wire 1.2 that fractured during the cyclic fatigue test at body temperature.

2.3. SEM Analysis

The fracture surfaces of all fragments were observed under a scanning electron microscope (SEM) (ZEISS Supra 35VP; GmBH, Oberkochen, Germany) for topographic features of the fractured instruments.

2.4. Statistical Analysis

Once the normality of distributions and equality of variances were confirmed by the Shapiro–Wilk and Levene tests, respectively, statistical analysis was performed using a two-way analysis of variance and Tukey's post hoc multiple comparison test, with NiTi wire dimension and heat treatment as the independent variables. The significance level was set at 5% ($p < 0.05$) using the statistical software Prism 8.0 (GraphPad Software, Inc., La Jolla, CA, USA).

3. Results

The means and standard deviations of cyclic fatigue resistance, torque maximum load, and angle of rotation until fracture for each instrument are shown in Table 1.

Table 1. Number of cycles to fracture [NCF] at room and body temperature, maximum torque [Ncm], and angle of rotation until fracture [°] values [mean ± standard deviation] of the different heat-treated TS2 instruments.

Instrument	Number of Cycles to Fracture [NCF]		Torque [Ncm]	Angle of Rotation [°]
	25 °C ± 1 °C	37 °C ± 1 °C		
	Mean ± SD	Mean ± SD	Mean ± SD	Mean ± SD
TS2 1.0 T-wire	395 [a1] ± 50	244 [b1] ± 42	0.41 [1] ± 0.06	294 [1] ± 42
TS2 1.2 T-wire	217 [a2] ± 48	142 [b2] ± 26	0.38 [1] ± 0.05	271 [1] ± 19
TS2 1.0 CM-wire	2153 [a3] ± 391	1608 [b3] ± 204	0.45 [1] ± 0.05	423 [2] ± 43
TS2 1.2 CM-wire	2303 [a3] ± 420	1825 [b3] ± 316	0.40 [1] ± 0.08	505 [2] ± 87

Same letters show differences not statistically significant between instruments in the same row ($p < 0.05$). Same numbers show differences not statistically significant between instruments in the same column ($p < 0.05$).

The TS2 CM-wire prototype files had the statistically highest cyclic fatigue resistance, at both temperatures, independently from the wire dimensions ($p < 0.05$).

When comparing T-wire files, TS2 mini 1.0 exhibited higher cyclic fatigue resistance than TS2 1.2 at both temperatures ($p < 0.05$). With regard to CM-wire files, no significant differences emerged between the two wire dimensions at room or body temperature ($p < 0.05$). Finally, being at body temperature reduced the fracture resistance of all instruments ($p < 0.05$).

The mean length of the fractured fragment (5 ± 0.2 mm) was not significantly different for the instruments tested ($p < 0.05$). No significant difference emerged in the maximum torque values between the tested instruments ($p < 0.05$). The CM-wire TS2 files showed significantly higher angular rotation to fracture than TS2 T-wire instruments ($p < 0.05$), with no significant difference between the two wire dimensions for both heat-treatments.

Scanning electron microscopy of the fracture surfaces exhibited similar and typical patterns of cyclic fatigue for all tested instruments. The crack initiation area and overload fast fracture zone for cyclic fatigue failure are shown in Figure 4.

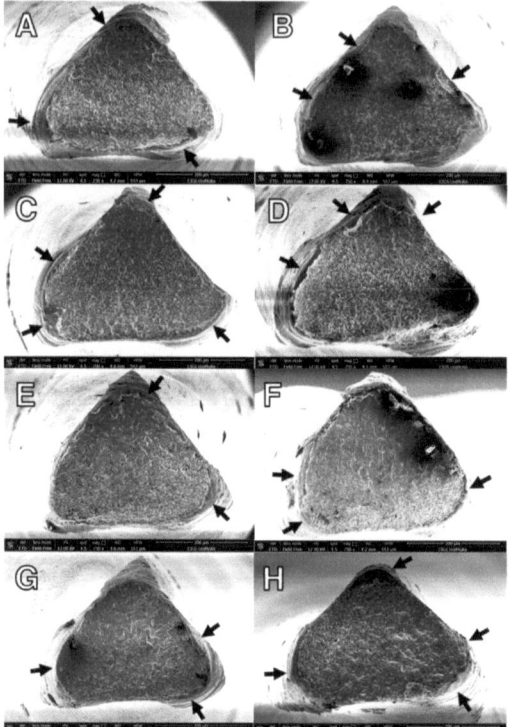

Figure 4. Representative field-emission scanning electron microscope images of the instruments after fatigue tests: (**A,B**) TS2 mini 1.0 T-wire, (**C,D**) TS2 1.2 T-wire, (**E,F**) TS2 mini 1.0 CM-wire, and (**G,H**) TS2 1.2 CM-wire, at 25 °C ± 1 °C (**A,C,E,G**) and 37 °C ± 1 °C (**B,D,F,H**). Typical features of cyclic fatigue fracture of the ductile fatigue area with microdimples and cones are observed with the black arrows indicating the origins of crack initiation.

4. Discussion

The current study aimed to assess the influence of NiTi wire dimensions on the mechanical properties of differently heat-treated TS2 prototypes. Although natural teeth best represent clinical conditions, variation in root canal anatomy is difficult to standardize in laboratory conditions [1]. Therefore, a stainless-steel canal, as previously described [15], was used for cyclic fatigue tests [15]. Moreover, the customized device ensured the reproducibility of the tests [15–17]. The correct placement of each file into the artificial canal was confirmed by the similar lengths of the fractured segments from the tested instruments [9]. Torsional tests were performed following the ISO Standard 3630-1, as previously reported [19].

According to the results of the present study, wire dimension had no significant effect for TS2 prototypes produced with CM heat treatment, which were significantly more resistant than TS2 prototypes with T-wire heat treatment, at both temperatures, independent of NiTi wire dimension. Conversely, the TS2 mini 1.0 T-wire showed higher cyclic fatigue resistance than the TS2 1.2 T-wire, at both temperatures. Thus, the first null hypothesis could be partially rejected. The results obtained for CM-wire files could be attributed to the heat treatment itself, which confers the rotary files more flexibility, as previously reported [3,20], compensating for the difference in the diameter dimensions of the NiTi wire of the two instruments [21,22]. The advantage provided by the T-wire heat treatment is less than the CM-wire [5,20,23], and consequently, the benefit of the smaller amount of metal in the TS2 mini 1.0 files is significantly more pronounced in the T-wire files in terms of increased flexibility and resistance to cyclic fatigue [3].

Moreover, our results showed that body temperature reduced the fracture resistance of all the instruments tested. These results support previous studies that found a reduction in the cyclic fatigue resistance of NiTi rotary instruments exposed at body temperature, due to changes in crystalline phases induced by the temperature increase [24–26].

Torsional tests showed that wire dimension had no significant effect on either maximum torque or angle of rotation to fracture. Thus, the second null hypothesis could not be rejected. The CM files exhibited significantly higher values of angular rotation than the T-wire files, whereas no significant difference emerged in terms of maximum torque values. The higher angular rotation (a parameter associated with ductile fractures) [19,27] could be attributable to the major flexibility of the CM-treatment [28], which does not impair maximum torque as confirmed by the similar values obtained with T-Wire files. In addition, the lack of a significant difference between the two wire dimensions suggest that torsional behavior is probably affected more by heat-treatment than by wire diameter. Since no authors have evaluated the mechanical properties of TS2 instruments with different NiTi wire diameters, some of our results cannot be confirmed by different studies.

SEM analysis revealed that fracture surfaces exhibited similar typical patterns of cyclic fatigue for all tested instruments. Of note, fractographic analysis of 1.0 T-wire files revealed a wider area covered by fatigue striations than in 1.2 T-wire instruments, thus suggesting slower fracture propagation after surface microcracks formed and, therefore, a higher resistance [29]. This analysis further corroborates the improvement in cyclic fatigue resistance of 2Shape 1.0 T-wire instruments when compared with their 1.2 counterparts.

Although the customized device we used was able to control all laboratory variables, we still faced some limitations. The artificial canal was specifically designed for the TS2 instrument employed, in terms of size (25) and taper (0.06), to ensure standardized conditions for both 0.06 tapered files. Yet, differences in file dimension could influence the adaptation of the file into the artificial walls and should be considered for future research. In clinical practice, many other factors, such as operator experience and anatomical complexities, act simultaneously, affecting the mechanical behavior of files used [30,31]. In addition, differences in the mechanical behavior of tested instruments should be determined, considering multiple factors including the phase transformation temperatures of heat-treated alloys [32]. CM alloys have been extensively investigated [28], whereas no data are available on T-wire phase transformation temperature. Future studies should address this matter. Furthermore, all variables associated with different NiTi endodontic files, such as design and dimensions, could impact their fracture resistance and need to be explored.

A previous study investigated the influence of environmental temperature on the torsional behavior of 25.06 conventional NiTi alloy and CM thermal-treated NiTi instruments, reporting no significant differences between files tested at 21 and 35 °C [33]. Thus, the authors hypothesized that variations in temperature would not affect torsional behavior. However, the differences in methodological conditions and lack of other findings require further investigations to confirm this hypothesis.

Within these limits, the present study has relevant clinical implications: heat treatments combined with a different NiTi wire dimensions (and in turn, different amounts of metal) could significantly affect the mechanical behavior of files. Limited to these laboratory conditions: CM-wires exhibited higher flexibility independently of wire dimensions, maintaining suitable torsional behavior; the reduction in NiTi wire dimension from 1.2 (TS2) to 1.0 (TS2 mini) increased fatigue resistance for T-wire heat-treated TS2. Clinicians should choose flexible instruments when flexural stress is high, such as in accentuate curvatures.

Future studies should investigate how the effects of NiTi wire dimensions combined with different heat treatments and file designs affect the mechanical properties of NiTi instruments.

5. Conclusions

Within the limits of the present study, the TS2 prototypes made by controlled-memory alloy showed higher cyclic fatigue resistance than conventional TS2 T-wires, independent of wire dimensions and test temperatures, exhibiting suitable torsional properties. In addition, the 1.0 wire diameter positively affected the cyclic fatigue resistance of TS2 mini T-wires, at both tested temperatures, whereas it had no effect on torsional behavior of CM-wire or T-wire heat-treated instruments.

Author Contributions: Conceptualization, E.P.; methodology, G.R.M.L.R. and F.S.C.; software, L.G.; validation, W.N., A.N., and F.D.; formal analysis, F.S.C. and L.G.; investigation, E.P.; resources, L.G.; data curation, G.R.M.L.R.; writing—original draft preparation, G.R.M.L.R. and F.S.C.; writing—review and editing, E.P., L.G., and W.N.; visualization, F.D. and A.N.; supervision, W.N.; project administration, E.P. All authors have read and agreed to the published version of the manuscript.

Funding: This research received no external funding.

Institutional Review Board Statement: Not applicable.

Informed Consent Statement: Not applicable.

Data Availability Statement: The data presented in this study are available on request from the corresponding author.

Conflicts of Interest: The authors declare no conflict of interest.

References

1. Özyürek, T.; Gündoğar, M.; Uslu, G.; Yılmaz, K.; Staffoli, S.; Nm, G.; Plotino, G.; Polimeni, A. Cyclic fatigue resistances of Hyflex EDM, WaveOne gold, Reciproc blue and 2shape NiTi rotary files in different artificial canals. *Odontology* **2018**, *106*, 408–413. [CrossRef] [PubMed]
2. Almeida, G.C.; Guimarães, L.C.; Resende, P.D.; Buono, V.; Peixoto, I.; Viana, A. Torsional behaviour of Reciproc and Reciproc blue instruments associated with their martensitic transformation temperatures. *Int. Endod. J.* **2019**, *52*, 1768–1772. [CrossRef] [PubMed]
3. Uslu, G.; Gundogar, M.; Özyurek, T.; Plotino, G. Cyclic fatigue resistance of reduced-taper nickel-titanium (NiTi) instruments in doubled-curved (S-shaped) canals at body temperature. *J. Dent. Res. Dent. Clin. Dent. Prospect.* **2020**, *14*, 111–115. [CrossRef]
4. Pedullà, E.; Lo Savio, F.; Boninelli, S.; Plotino, G.; Grande, N.M.; Rapisarda, E.; La Rosa, G. Influence of cyclic torsional preloading on cyclic fatigue resistance of nickel—titanium instruments. *Int. Endod. J.* **2015**, *48*, 1043–1050. [CrossRef]
5. Pedullà, E.; La Rosa, G.; Virgillito, C.; Rapisarda, E.; Kim, H.C.; Generali, L. Cyclic Fatigue Resistance of Nickel-titanium Rotary Instruments according to the Angle of File Access and Radius of Root Canal. *J. Endod.* **2020**, *46*, 431–436. [CrossRef] [PubMed]
6. Arias, A.; Hejlawy, S.; Murphy, S.; de la Macorra, J.C.; Govindjee, S.; Peters, O.A. Variable impact by ambient temperature on fatigue resistance of heat-treated nickel titanium instruments. *Clin. Oral Investig.* **2019**, *23*, 1101–1108. [CrossRef]
7. de Vasconcelos, R.A.; Murphy, S.; Carvalho, C.A.; Govindjee, R.G.; Govindjee, S.; Peters, O.A. Evidence for Reduced Fatigue Resistance of Contemporary Rotary Instruments Exposed to Body Temperature. *J. Endod.* **2016**, *42*, 782–787. [CrossRef]
8. Jamleh, A.; Yahata, Y.; Ebihara, A.; Atmeh, A.R.; Bakhsh, T.; Suda, H. Performance of NiTi endodontic instrument under different temperatures. *Odontology* **2016**, *104*, 324–328. [CrossRef]
9. Gündoğar, M.; Özyürek, T. Cyclic Fatigue Resistance of OneShape, HyFlex EDM, WaveOne Gold, and Reciproc Blue Nickel-titanium Instruments. *J. Endod.* **2017**, *43*, 1192–1196. [CrossRef]
10. Gündoğar, M.; Özyürek, T.; Yılmaz, K.; Uslu, G. Cyclic fatigue resistance of HyFlex EDM, Reciproc Blue, WaveOne Gold, and Twisted File Adaptive rotary files under different temperatures and ambient conditions. *J. Dent. Res. Dent. Clin. Dent. Prospect.* **2019**, *13*, 166–171. [CrossRef]
11. Pereira, E.S.; Gomes, R.O.; Leroy, A.M.; Singh, R.; Peters, O.A.; Bahia, M.G.; Buono, V.T. Mechanical behavior of M-Wire and conventional NiTi wire used to manufacture rotary endodontic instruments. *Dent. Mater.* **2013**, *29*, e318–e324. [CrossRef] [PubMed]
12. Elnaghy, A.M.; Elsaka, S.E. Torsional resistance of XP-endo Shaper at body temperature compared with several nickel-titanium rotary instruments. *Int. Endod. J.* **2018**, *51*, 572–576. [CrossRef] [PubMed]
13. Adiguzel, M.; Isken, I.; Pamukcu, I.I. Comparison of cyclic fatigue resistance of XP-endo Shaper, HyFlex CM, FlexMaster and Race instruments. *J. Dent. Res. Dent. Clin. Dent. Prospect.* **2018**, *12*, 208–212. [CrossRef]
14. Ataya, M.; Ha, J.H.; Kwak, S.W.; Abu-Tahun, I.H.; El Abed, R.; Kim, H.C. Mechanical Properties of Orifice Preflaring Nickel-titanium Rotary Instrument Heat Treated Using T-Wire Technology. *J. Endod.* **2018**, *44*, 1867–1871. [CrossRef]
15. Pedullà, E.; La Rosa, G.; Boninelli, S.; Rinaldi, O.G.; Rapisarda, E.; Kim, H.C. Influence of Different Angles of File Access on Cyclic Fatigue Resistance of Reciproc and Reciproc Blue Instruments. *J. Endod.* **2018**, *44*, 1849–1855. [CrossRef] [PubMed]

16. La Rosa, G.; Palermo, C.; Ferlito, S.; Isola, G.; Indelicato, F.; Pedullà, E. Influence of surrounding temperature and angle of file access on cyclic fatigue resistance of two single file nickel-titanium instruments. *Aust. Endod. J.* **2021**, *47*, 260–264. [CrossRef] [PubMed]
17. La Rosa, G.; Shumakova, V.; Isola, G.; Indelicato, F.; Bugea, C.; Pedullà, E. Evaluation of the Cyclic Fatigue of Two Single Files at Body and Room Temperature with Different Radii of Curvature. *Materials* **2021**, *14*, 2256. [CrossRef]
18. Thu, M.; Ebihara, A.; Maki, K.; Miki, N.; Okiji, T. Cyclic Fatigue Resistance of Rotary and Reciprocating Nickel-Titanium Instruments Subjected to Static and Dynamic Tests. *J. Endod.* **2020**, *46*, 1752–1757. [CrossRef]
19. Pedullà, E.; LA Rosa, G.; Franciosi, G.; Corsentino, G.; Rapisarda, S.; Lo Savio, F.; La Rosa, G.; Grandini, S. Cyclic fatigue and torsional resistance evaluation of Reciproc R25 instruments after simulated clinical use. *Minerva Dent. Oral Sci.* **2022**, *71*, 174–179. [CrossRef]
20. Gündoğar, M.; Uslu, G.; Özyürek, T.; Plotino, G. Comparison of the cyclic fatigue resistance of VDW.ROTATE, TruNatomy, 2Shape, and HyFlex CM nickel-titanium rotary files at body temperature. *Restor. Dent. Endod.* **2020**, *45*, e37. [CrossRef]
21. Zhou, H.M.; Shen, Y.; Zheng, W.; Li, L.; Zheng, Y.F.; Haapasalo, M. Mechanical properties of controlled memory and superelastic nickel-titanium wires used in the manufacture of rotary endodontic instruments. *J. Endod.* **2012**, *38*, 1535–1540. [CrossRef]
22. Goo, H.J.; Kwak, S.W.; Ha, J.H.; Pedullà, E.; Kim, H.C. Mechanical Properties of Various Heat-treated Nickel-titanium Rotary Instruments. *J. Endod.* **2017**, *43*, 1872–1877. [CrossRef] [PubMed]
23. Koçak, S.; Şahin, F.F.; Özdemir, O.; Koçak, M.M.; Sağlam, B.C. A comparative investigation between ProTaper Next, Hyflex CM, 2Shape, and TF-Adaptive file systems concerning cyclic fatigue resistance. *J. Dent. Res. Dent. Clin. Dent. Prospect.* **2021**, *15*, 172–177. [CrossRef]
24. Grande, N.M.; Plotino, G.; Silla, E.; Pedullà, E.; DeDeus, G.; Gambarini, G.; Somma, F. Environmental Temperature Drastically Affects Flexural Fatigue Resistance of Nickel-titanium Rotary Files. *J. Endod.* **2017**, *43*, 1157–1160. [CrossRef] [PubMed]
25. Klymus, M.E.; Alcalde, M.P.; Vivan, R.R.; Só, M.; de Vasconselos, B.C.; Duarte, M. Effect of temperature on the cyclic fatigue resistance of thermally treated reciprocating instruments. *Clin. Oral Investig.* **2019**, *23*, 3047–3052. [CrossRef] [PubMed]
26. Dosanjh, A.; Paurazas, S.; Askar, M. The Effect of Temperature on Cyclic Fatigue of Nickel-titanium Rotary Endodontic Instruments. *J. Endod.* **2017**, *43*, 823–826. [CrossRef] [PubMed]
27. Weissheimer, T.; Heck, L.; Calefi, P.; Alcalde, M.P.; da Rosa, R.A.; Vivan, R.R.; Duarte, M.; Só, M. Evaluation of the mechanical properties of different nickel-titanium retreatment instruments. *Aust. Endod. J.* **2021**, *47*, 265–272. [CrossRef] [PubMed]
28. Zupanc, J.; Vahdat-Pajouh, N.; Schäfer, E. New thermomechanically treated NiTi alloys—A review. *Int. Endod. J.* **2018**, *51*, 1088–1103. [CrossRef]
29. Azizi, A.; Prati, C.; Schiavon, R.; Fitzgibbon, R.M.; Pirani, C.; Iacono, F.; Pelliccioni, G.A.; Spinelli, A.; Zamparini, F.; Puddu, P.; et al. In-depth metallurgical and microstructural analysis of Oneshape and heat treated Onecurve instruments. *Eur. Endod. J.* **2021**, *6*, 90–97. [CrossRef]
30. Zanza, A.; D'Angelo, M.; Reda, R.; Gambarini, G.; Testarelli, L.; Di Nardo, D. An Update on Nickel-Titanium Rotary Instruments in Endodontics: Mechanical Characteristics, Testing and Future Perspective-An Overview. *Bioengineering* **2021**, *8*, 218. [CrossRef]
31. Plotino, G.; Grande, N.M.; Cordaro, M.; Testarelli, L.; Gambarini, G. A review of cyclic fatigue testing of nickel-titanium rotary instruments. *J. Endod.* **2009**, *35*, 1469–1476. [CrossRef]
32. Silva, E.J.N.L.; Martins, J.N.R.; Ajuz, N.C.; Antunes, H.S.; Vieira, V.T.L.; Braz Fernandes, F.M.; Belladonna, F.G.; Versiani, M.A. A Multimethod assessment of a new customized heat-treated nickel–titanium rotary file system. *Materials* **2022**, *15*, 5288. [CrossRef]
33. Silva, E.; Giraldes, J.; de Lima, C.O.; Vieira, V.; Elias, C.N.; Antunes, H.S. Influence of heat treatment on torsional resistance and surface roughness of nickel-titanium instruments. *Int. Endod. J.* **2019**, *52*, 1645–1651. [CrossRef]

Article

Comparing Cyclic Fatigue Resistance and Free Recovery Transformation Temperature of NiTi Endodontic Single-File Systems Using a Novel Testing Setup

Emad Youssef [1,2,*], Holger Jungbluth [1], Søren Jepsen [1], Manfred Gruener [2] and Christoph Bourauel [2]

1 Department of Periodontology, Operative and Preventive Dentistry, University of Bonn, 53111 Bonn, Germany
2 Department of Oral Technology, Faculty of Medicine, University of Bonn, 53111 Bonn, Germany
* Correspondence: e.youssef@web.de; Tel.: +49-0228-287-22428

Citation: Youssef, E.; Jungbluth, H.; Jepsen, S.; Gruener, M.; Bourauel, C. Comparing Cyclic Fatigue Resistance and Free Recovery Transformation Temperature of NiTi Endodontic Single-File Systems Using a Novel Testing Setup. *Materials* **2024**, *17*, 566. https://doi.org/10.3390/ma17030566

Academic Editors: Yeong-Joon Park and Rui Miranda Guedes

Received: 11 December 2023
Revised: 5 January 2024
Accepted: 22 January 2024
Published: 25 January 2024

Copyright: © 2024 by the authors. Licensee MDPI, Basel, Switzerland. This article is an open access article distributed under the terms and conditions of the Creative Commons Attribution (CC BY) license (https:// creativecommons.org/licenses/by/ 4.0/).

Abstract: The aim of this study was to assess the effect of body temperature (37 °C) on the cyclic fatigue resistance of three endodontic single-file systems using a new testing setup. One Shape® new generation (OS), WaveOne™ (WO) and WaveOne® GOLD (WOG), which are made from different NiTi alloys and operated in different motions (rotation/reciprocation), were evaluated. The study design included four groups. Each group comprised 30 files, 10 files of each of the three file systems, tested at 20 ± 2 °C (group 1 and 3) and at 37 ± 1 °C (group 2 and 4). All files were tested in a custom-made metal block with artificial canals of 60° angle, and a 5 mm and 3 mm radius of curvature, respectively. A heating element was attached to replicate a temperature of 37 °C. Files were introduced 18 mm into the canals and operated until failure. Transformation temperatures of five samples of each of the tested file systems were determined via the bend and free recovery (BFR) method. With the exception of WOG in canals with a 3 mm radius of curvature ($p = 0.075$), all the tested file systems showed statistically significantly less time needed to fracture when operated at 37 ± 1 °C compared to at 20 ± 2 °C in canals with a 5 mm and 3 mm radius of curvature using Mann–Whitney U test ($p < 0.05$). All file systems showed transformation temperatures below the body temperature. We concluded that body temperature directly affects the cyclic fatigue resistance of all tested file systems. Bend and free recovery can be suitable for the determination of austenite finish temperatures (A_f) of endodontic instruments as it allows testing a longer portion of the instrument.

Keywords: bend and free recovery; cyclic fatigue; reciprocating endodontic instruments

1. Introduction

Root canal instrumentation requires the use of highly flexible instruments that can easily bend to clean and shape curved root canals while respecting the original canal anatomy. Due to its low modulus of elasticity, nickel-titanium (NiTi) alloy gained interest in being used in fabricating root canal instruments in the last three decades.

In 1988, Walia et al. were the first to report that K-files #15 manufactured from NiTi alloy showed two to three times more elastic flexibility in bending and in torsion in addition to higher torsional fracture resistance compared to stainless steel counterparts that had the same size and the same manufacturing process [1].

In 1992, the first generation of rotary 0.02 taper NiTi files came to the market and its introduction revolutionized the practice of endodontics and changed the way of mechanical canal instrumentation [2]. Engine-driven rotary canal instrumentation allowed faster canal preparation and minimized procedural errors associated with hand instrumentation, which was advantageous for both experienced and inexperienced operators [3].

Despite its many advantages, a major concern with using engine-driven NiTi instruments was their liability to fracture [4,5]. Separation of engine-driven NiTi instruments can occur due to either torsional fracture (shear failure) or cyclic fatigue (flexural fatigue) or a combination of both [4,6]. Many NiTi file systems were developed to reduce the incidence of instrument separation inside root canals. The major changes made were in terms of instrument design, type of cutting motion inside the root canal (full rotation and reciprocation), alloy of the instrument and method of instrument fabrication (heat treatment and electrical discharge machining).

In order to better understand the reasons for instrument separation, many in vitro studies were conducted comparing the behavior of different file systems in standardized artificial canals with different radii and angles of curvature [7–11]. Not until recently, all of the available in vitro studies on cyclic fatigue were conducted at room temperature. NiTi is considered to be a shape memory alloy that exhibits pseudoelasticity (super-elasticity) and the shape memory effect [12]. The shape memory effect refers to the ability of a material to restore its original shape after heating. This indicates that NiTi can be sensitive to temperature changes and shows different behaviors in different temperatures due to different specific heat capacities of different material phases [13].

Most of the engine-driven root canal instruments are made of 55 Nitinol alloy [≈56% (wt) nickel and 44% (wt) titanium] and characteristically have two phases, an austenitic phase where the crystal structure of the alloy is a stable, body-centered cubic lattice, and a martensitic phase, with the crystal structure of a closely packed hexagonal lattice [14]. Temperature or stress-induced change between these two phases can occur, resulting in the previously mentioned characteristics termed shape memory and super-elasticity.

When an NiTi instrument is used to clean and shape a curved root canal in a patient's tooth, it is subjected to two different hysteresis, stress-induced martensitic transformation (due to canal curvature) and, simultaneously, heat-induced austenitic transformation (due to body temperature). Therefore, in our study we wanted to reproduce these conditions of files operated at body temperature in comparison to room temperature in curved canals.

In this study, we compared the cyclic fatigue resistance of three single-file systems, One Shape® new generation (OS), WaveOne™ (WO), WaveOne® GOLD (WOG), which have the same tip diameter (iso #25), made from different alloys and operated in different motions (rotation/reciprocation). Bend and free recovery testing was then used to assess the transformation temperature of the three file systems included in this study.

The null hypothesis is that there is no difference in cyclic fatigue characteristics of NiTi instruments, when driven at room temperature or at body temperature.

2. Materials and Methods

In this study, a total of 120 files divided into 4 groups were included. Each group comprised 30 files, 10 files from each of the tested three file systems: One Shape® new generation (OS) (MICRO-MEGA, Besançon, France), WaveOne™ Primary (WO) (Dentsply Maillefer, Ballaigues, Switzerland) and WaveOne® GOLD primary (WOG) (Dentsply Maillefer, Ballaigues, Switzerland). Figure 1 shows the groups distribution.

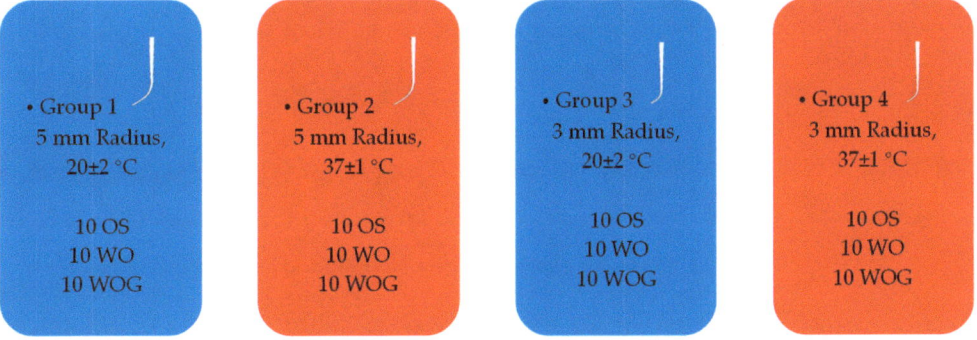

Figure 1. Groups distribution.

In Group 1 (G1) and Group 3 (G3), the files were tested in a metal block with artificial canals of a 5 mm and 3 mm radius of curvature at room temperature 20 ± 2 °C. In Group 2 (G2) and Group 4 (G4), the files were tested in artificial canals with a 5 mm and 3 mm radius of curvature at simulation of body temperature of 37 ± 1 °C. In all groups, the artificial canal angle of curvature was kept the same at 60° as determined by Pruett et al. [15].

2.1. Cyclic Fatigue Testing
2.1.1. Testing Setup

A self-developed mini universal testing machine was used (Figure 2). A handpiece holder was fixed to the upper end of the machine while the stainless steel block (manufacturing method is described in Appendix A) was attached to the lower part (moving part). The stainless steel block was fixed to a stainless steel holder that assured correct and reproducible positioning. The holder thickness was 8 mm and was equipped with a Peltier element (semiconductor-heating element) attached to it as shown in Figure 2C. The Peltier element was connected to a power control box for adjusting the current and voltage needed to generate a temperature of 37 ± 1 °C in the stainless steel block. The holder was fixed to the lower part (moving part) of the testing machine by means of an electromagnet connected to a force detecting sensor.

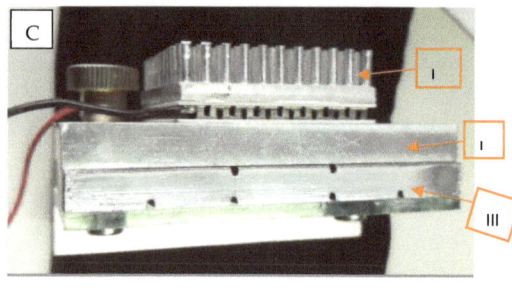

Figure 2. (**A**) Testing setup. (**B**) Testing block attached to L-shaped block holder which is kept in place in testing machine by means of electromagnet. Handpiece with tested file is mounted. (**C**) Peltier element (I) attached to L-shaped holder (II) and testing block (III) is fixed in place with a screw. (**D**) Thermometer metal probe (i) placed in a hole drilled in the testing block to measure core temperature of the block.

Using the metal holder allowed a uniform heat transfer to the stainless steel testing block and easy removal of the block, for retrieval of the separated instrument fragments, without changing the position relation between the handpiece and the block holder. Keeping a fixed relation between the handpiece holder and block holder allowed the reproducibility of file position in the canals after removing the separated fragment. The force sensor connected to the magnet holding the block holder ensured the correct insertion angle each time a new file was used. The force sensor readings showed that the files were introduced in the canal freely (at $0°$ inclination) until first contact with the canal curvature as inclined insertion angle can affect the cyclic fatigue of tested instruments [16].

The universal testing machine was connected to a computer where the force sensor readings can be monitored, and through a self-developed controlling program, the introduction of the files in the canals was controlled. The files were introduced 18 mm inside the canals after lubricating the canals with non-CFC propellant (KaVo Spray™, KaVo Dental GmbH, Biberach an der Riss, Germany). A camera setup (S5700 Fujifilm Group, Tokyo, Japan) was placed directly in front of the glass cover and an FHD video recording was started, and was used later for confirming the time to fracture. A metal probe thermometer (AZ 8856, Taichung, Taiwan) was placed in a 7 mm deep hole drilled on the side of the stainless steel testing block for continuous monitoring of temperature during testing (Figure 2D). The 6:1 reduction handpiece (Sirona Dental Systems GmbH, Bensheim, Germany) was connected to an endomotor (Reciproc Gold, VDW, Munich, Germany) and the operating mode was chosen according to the manufacturer's instructions. For WOG and WO, the mode "Waveone All" was chosen, where both the instruments were operated in a partial reciprocation movement with a forward angle of rotation of $150°$ and a reverse angle of $30°$. For OS, the speed was set to 350 rpm and the torque was set to 2.5 N.cm.

2.1.2. Testing Procedure

The rotation/reciprocation of files in artificial canals was controlled through the foot pedal and stopwatch. Operation was ceased upon visual or auditory separation detection. The retrieval process included lowering the block holder with the block, unscrewing the stainless steel block and storing the separated fragment along with the rest of the tested file. After attaching a new file to the handpiece and fixing the block to its holder, the block and its holder were moved up as an assembly until the new file was introduced 18 mm in the artificial canal to start a new test cycle. Tests of G1 and G3 were conducted in an air-conditioned room with a fixed temperature set to prevent temperature changes. The setup was left overnight in the room and temperature was recorded by the thermometer attached to the testing block during each test. The recorded temperatures were found to be in the range of 20 ± 2 °C. For G2 and G4, the temperature was continuously monitored and recorded during the testing procedure and the current intensity was adjusted (if needed) simultaneously to maintain the real-time temperature in the range of 37 ± 1 °C.

2.2. Assessment of Transformation Temperature of NiTi Using Bend and Free Recovery Test

A self-developed bend and free recovery (BFR) testing machine was used (Figure 3). The machine was built following the ASTM international standard "Standard test method for determination of transformation temperature of Nickel-Titanium shape memory alloys by bend and free recovery" (Designation: ASTM F2082-03) [17]. The machine consists of a fluid pump with a thermostat, double-chambered glass container, displacement laser sensor, sample holder, carbon fiber rod, thermometer and a computer.

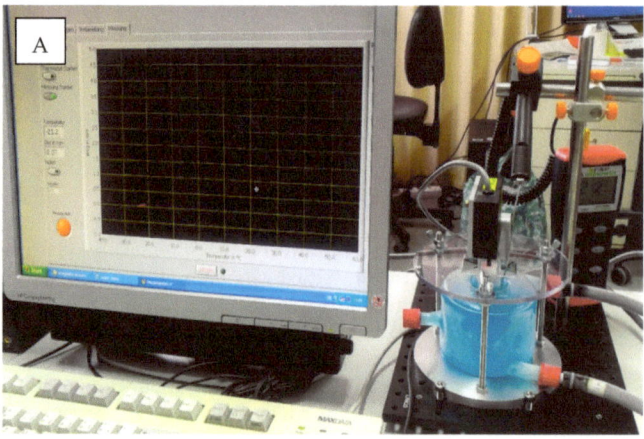

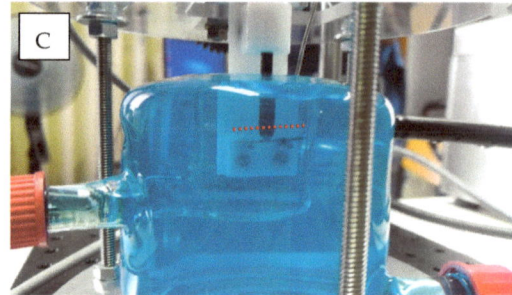

Figure 3. (**A**) Bend and free recovery testing setup showing the double chamber glass, laser sensor, thermometer and the connected computer for real-time plotting of displacement/temperature changes. (**B**) Closer look at the bent sample. (**C**) Straightening of the sample after reaching active A_f temperature. At both (**B**,**C**), a red dotted line parallel to the sample is drawn for illustration.

Three groups (*n* = 5) representing the three used file systems were included. Samples of 20 mm (from the tip) were prepared by cutting the shaft of each file with a diamond disk. Laser sensor, thermometer and thermostat were started, all connections to the computer were checked and the measuring software was started on the computer. On the double-pin sample holder, which is placed inside the smaller chamber of the glass container, a 20 mm file sample was placed. The carbon rod was then seated on the straight sample nearly in the middle of the sample (Figure 3) and the thermometer was placed inside the smaller chamber not touching the sample.

Four liters of cooling liquid, pre-cooled at -20 °C for 24 h and at -80 °C for 3 h, were then poured into the pump and the pump was started. When the liquid filled the bigger chamber in the glass container, the cooling liquid was filled carefully inside the smaller chamber using a 20 mL syringe. The carbon rod was then pressed 2–4 mm down while resting on the cooled sample (martensite phase), which resulted in curving the sample with max. point of curvature near the mid-length (Figure 3B). The laser sensor was then focused on the carbon rod end and the zero position was recorded. When the coldest temperature was recorded, the measurement cycle was started. The pump was previously set to heat the liquid that was pumped to the bigger chamber of the glass container. The liquid in the smaller chamber was heated by heat conduction and the temperature was recorded continuously.

The rise in temperature of the liquid in contact with the sample resulted in a phase transformation of the NiTi and caused the sample to straighten back to its original straight form. When the sample straightened back, it caused the freely moving carbon rod resting on the sample to move upwards and the laser sensor recorded this displacement. The recorded displacement with the continuously recorded temperature data was used to plot a graph that shows the transformation temperature due to free recovery of the tested sample.

2.3. Scanning Electron Microscopy

The fractured segments underwent analysis with a scanning electron microscope (XL30 SEM, Philips, Eindhoven, The Netherlands) to validate that the fracture occurred as a result of cyclic fatigue, displaying the characteristic cyclic fatigue fracture pattern.

2.4. Statistical Analyses

The assumption of normal distribution of data was tested using the Shapiro–Wilk test. The time to fracture data for each file system tested at two different ambient temperatures or in canals with two different radii were analyzed via the Mann–Whitney U test, while the analysis of time to fracture between different file systems within the same group was done using Mood's median test. IBM SPSS Statistics for Windows (Version 23.0., Armonk, NY, USA: IBM Corp) was used for all statistical analyses.

3. Results

3.1. Cyclic Fatigue Resistance

Table 1 shows the mean time to fracture (in seconds) and standard deviation for One Shape® new generation (OS), WaveOne™ (WO) and WaveOne® GOLD (WOG) at room temperature (20 ± 2 °C) and body temperature (37 ± 1 °C). When tested in artificial canals with a 5 mm radius of curvature, all tested file systems showed statistically significantly less time to fracture at body temperature compared to room temperature ($p < 0.05$). In canals with a 3 mm radius, both WO and OS showed the same results, while time to fracture at body temperature was statistically significant lower compared to when tested at room temperature ($p < 0.05$) (Figure 4). For WOG, the statistical difference was insignificant at the 95% confidence level ($p > 0.05$). At both, room temperature (20 ± 2 °C) and body temperature (37 ± 1 °C), both of OS and WO, showed statistically significantly less time to fracture when tested in artificial canals with a 3 mm radius compared to artificial canals with a 5 mm radius of curvature ($p < 0.05$). For WOG, the time needed to fracture in 5 mm canals was significantly longer than in 3 mm canals at room temperature ($p < 0.001$), but when tested at body temperature, WOG instruments showed no difference in time to fracture between 5 mm and 3 mm canals ($p = 0.353$).

Table 1. Mean time to fracture (in seconds).

	Mean Time to Fracture (Seconds)				p-Value	
	Group 1 (G1) (5 mm, Room temperature 20 ± 2 °C)	Group 2 (G2) (5 mm, Body temperature 37 ± 1 °C)	Group 3 (G3) (3 mm, Room temperature 20 ± 2 °C)	Group 4 (G4) (3 mm, Body temperature 37 ± 1 °C)	G1–G2	G3–G4
One Shape® new generation (OS)	52 ± 12	34 ± 5	40 ± 9	26 ± 6	<0.001	0.002
WaveOne™ (WO)	221 ± 28	135 ± 25	114 ± 21	81 ± 12	<0.001	<0.001
WaveOne® GOLD (WOG)	224 ± 14	122 ± 15	134 ± 30	115 ± 15	<0.001	0.075

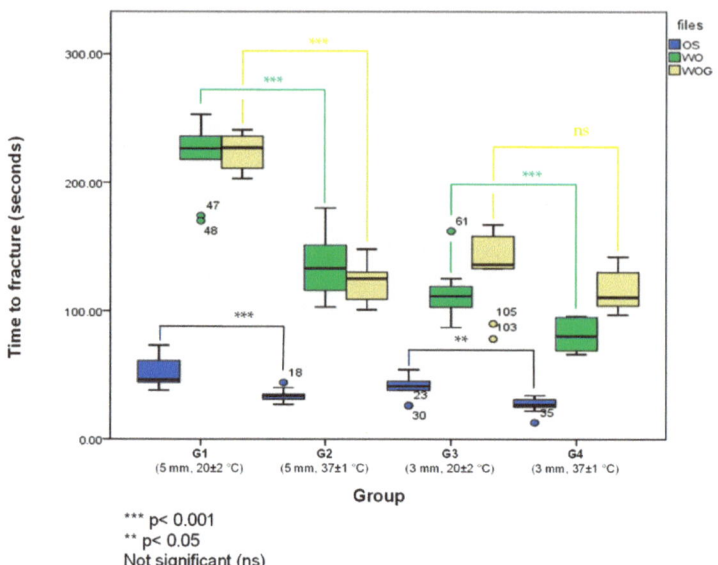

Figure 4. Mean time to fracture by groups. Group 1 (G1): 5 mm 20 ± 2 °C, Group 2 (G2): 5 mm 37 ± 1 °C, Group 3 (G3): 3 mm 20 ± 2 °C, Group 4 (G4): 3 mm 37 ± 1 °C.

Comparing the three file systems within the same group showed that OS (Rotating motion) had statistically significantly less time to fracture in all tested conditions compared to the two other file systems WO and WOG (reciprocating motion) ($p < 0.001$). Comparing the time to fracture of both reciprocation systems, WO and WOG, showed that in canals with a 5 mm radius of curvature, no statistical significance was detected at both testing temperatures ($p > 0.1$). When tested in canals with a 3 mm radius of curvature, WOG had statistically significantly longer time to fracture compared to WO at both testing temperatures ($p < 0.007$).

3.2. Transformation Temperature by Bend and Free Recovery

Table 2 shows the mean and standard deviation of active austenite start (A_s) and active austenite finish (A_f) temperatures as determined by the bend and free recovery test. WOG showed the highest A_s and A_f temperatures, while OS showed the lowest (Figure 5). All tested samples had an Af temperature below body temperature (37 ± 1 °C) (Figure 6).

Table 2. Mean active austenite start (A_s) and mean active austenite finish (A_f) temperatures as determined by bend and free recovery test.

	A_s (°C)	A_f (°C)
OS		
Mean	−17	−12
SD	3	2
WO		
Mean	−14	6
SD	3	5
WOG		
Mean	21	25
SD	3	2

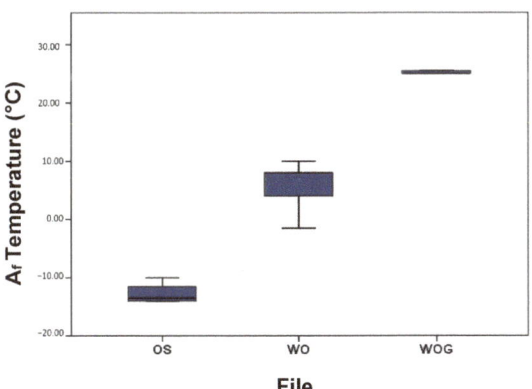

Figure 5. Mean active austenite finish (A_f) temperatures as determined by bend and free recovery test.

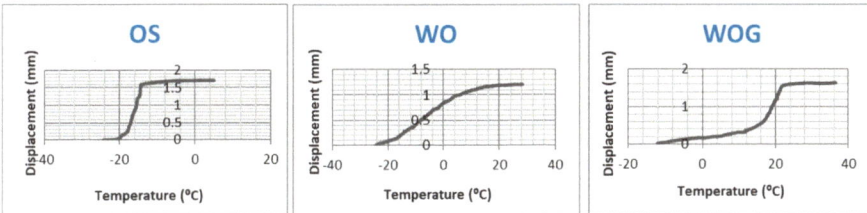

Figure 6. Active A_f Temperature of OS, WO and WOG as determined by bend and free recovery testing.

3.3. Scanning Electron Microscopy

Analysis of the cross-section of fractured surfaces using scanning electron microscopy (XL30 SEM, Philips, Eindhoven, The Netherlands) showed characteristic fatigue fracture patterns. As shown in Figure 7, areas of crack initiation (I), crack propagation (P) and areas of catastrophic fracture with ductile failure (dotted line) that showed prominent dimple defects are identifiable. Micro-voids (Figure 7 red arrows) and inclusions (cones) (Figure 7 yellow arrows) of different shapes and sizes were also observed distributed among the fracture surface with higher magnification. Fatigue striations (black arrows) could also be detected at higher magnification.

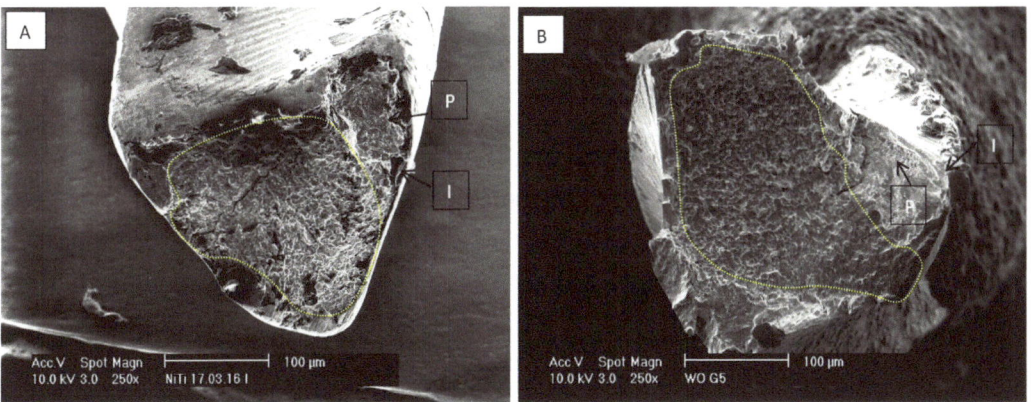

Figure 7. *Cont.*

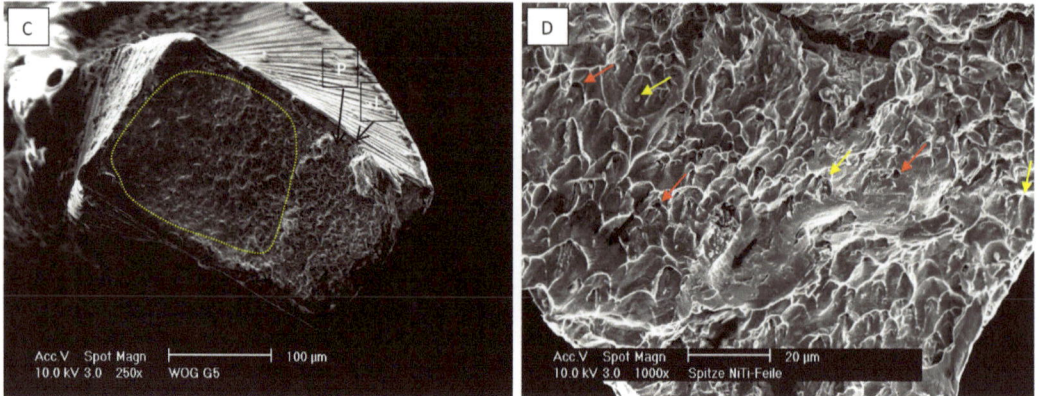

Figure 7. Scan electron microscope images of OS (**A**), WO (**B**) and WOG (**C**) where areas of crack initiation (I) and propagation (P) are identifiable. Area marked with dotted line shows catastrophic ductile failure. (**D**) Red arrows: micro-voids, yellow arrows: Inclusions/cones and black arrows: fatigue striations.

4. Discussion

New advancements are introduced continuously to present new engine-driven instruments with improved properties that can provide better performance in various clinical situations. One of these advancements was the introduction of partial reciprocating motion of endodontic files with rotation effect, where the instrument rotates in the canal with an angle of rotation in the forward cutting direction greater than the angle of rotation in the non-cutting opposite direction [18]. The instrument completes one turn after a certain number of reciprocating cycles [19]. Both file systems that were reciprocating (WO and WOG) showed significantly longer time to fracture in all tested groups when compared to OS, which was operated in full rotation. This confirms the findings of previous studies that showed that reciprocating instruments needed significantly longer time to failure compared to the full rotation instruments [20–25]. These results can be attributed to the fact that reciprocating instruments need more time to complete the same number of cycles as full rotation instruments or they are operated at a lower rpm.

When NiTi instruments are introduced into a curved canal, stress exerted from the curvature causes a stress-induced martensite transformation within the instrument. Thermal-induced phase transformation of NiTi is a known feature of these alloys, where a martensite to austenite transformation happens upon heating [14]. In 2016, body temperature was recognized as a factor that can alter the cyclic fatigue resistance of endodontic file systems [26]. Our results showed that when tested in moderately curved canals with a 5 mm radius of curvature, all tested file systems needed significantly less time to fracture at body temperature compared to room temperature ($p < 0.05$). In canals with a 3 mm radius, both WO and OS showed the same results, where time to fracture at body temperature was statistically significant lower compared to when tested at room temperature ($p < 0.05$). These findings confirm the results of previous studies reporting that environmental temperature has an influence on the cyclic fatigue resistance of NiTi-shaping instruments [26–31]. For WOG, the statistical difference was insignificant at 95% confidence ($p > 0.05$). Although no statistically significant difference was found for WOG when tested in canals with a 3 mm radius, the mean time to fracture at body temperature (115 ± 15 s) was still shorter than at room temperature (134 ± 30 s). This finding could be attributed to the sample size ($n = 10$) and a statistical significance might be detected with a larger sample size.

In 1997, Pruett et al. reported that there was an inverse relation between the radius of canal curvature and the cyclic fatigue resistance of NiTi instruments [15]. In our study, canals with 3 mm and 5 mm radii of curvature were used to represent canals with severe and moderate curvatures [32] that can be difficult to manage clinically. Our results showed that all tested file systems needed statistically significantly less time to fracture when tested in artificial canals with a 3 mm radius of curvature compared to artificial canals with a 5 mm radius at room temperature (20 ± 2 °C). This is consistent with previous findings in the literature [15,16,33,34]. At body temperature (37 ± 1 °C), both OS and WO showed the same findings, where time to fracture was significantly lower in 3 mm than in 5 mm canals ($p < 0.005$); however, WOG showed different results, where no significant difference could be observed between the time to fracture in 3 mm and 5 mm canals.

Resistance to cyclic fatigue can be influenced by various factors including the instrument design (taper, cross-section, central core mass, etc.) and alloy of the instrument [35]. The tested file systems in our study have the same tip diameter but they have different cross-sectional geometries. The OS cross-section at the apical portion is a modified triangle with a symmetrical radius and three cutting edges, the middle of the instrument has a transition to two cutting edges and the coronal portion has an S-shaped cross-section with two cutting edges. WO has a convex triangular cross-section coronally and a concave triangle at the apical portion. WOG has a parallelogram-shaped cross-section with two cutting edges in contact with the canal wall, alternating with an off-centered cross-section where only one cutting edge is in contact with the canal wall [36]. OS has a constant taper of 6% along the cutting part of the instrument, while both WO and WOG are designed to have a constant taper of 8% and 7%, respectively, for the most apical 3 mm. Then both instruments have a variable/regressive taper until the end of their cutting portion. The used instruments OS, WO and WOG were made of conventional super-elastic NiTi, M-wire and G-wire, respectively. When we compared the time to fracture of WO and WOG tested at the same group, we observed no significant difference in moderately curved canals (5 mm) ($p > 0.05$) but in severely curved canals (3 mm), WOG needed significantly longer time to fracture than WO ($p < 0.05$). This shows that WOG had a superior performance when compared to WO in severely curved canals.

For testing the cyclic fatigue of engine-driven file systems, many studies used a setup where the files were immersed in a water or saline bath that was heated to body temperature (37 ± 1 °C) [37–40]. Yum et al. [41] explained the galvanic corrosion that happened to Protaper Universal (PT) (Dentsply Maillefer, Ballaigues, Switzerland) instruments. They reported that EDX micro analysis of PT instruments revealed that the cutting and non-cutting sections of the instruments were made of NiTi, while the shank was made of gold-plated Brass and when in a galvanic cell, the NiTi acts as the anode and the shank acts as the cathode. Although the effect of galvanic corrosion can be insignificant due to the short testing duration (minutes), we used a new setup where a Peltier heating element attached to metal holder transferred the heat to the testing block. The metal probe thermometer placed in a hole drilled directly in the testing block served for accurate real-time monitoring of the core temperature in the testing block.

The bend and free recovery (BFR) test and differential scanning calorimetry (DSC) are two methods used for measuring the transformation temperatures of nickel-titanium (NiTi) shape memory alloys [42]. The DSC monitors the heat flow during both cooling and heating, whereas BFR records the deflection recovery only during heating [42]. The key transformation temperatures austenite start (A_s) and austenite finish (A_f) can be determined using either method. BFR can be performed directly on finished products and simulates the actual conditions that a product will experience in use, providing a more realistic evaluation of its performance. The bend and free recovery (BFR) method was used to determine the active austenite finish temperature (A_f) of the instruments. Using this method allowed us to test a 20 mm fragment of the instrument including the whole cutting section of the instrument (16 mm), unlike the case when differential scanning calorimetry (DSC) was used, where a small sample (usually 5 mm from the tip portion of the instrument) was used to determine the martensite–austenite transformation temperatures. All of the tested file systems showed A_f below body temperature (37 ± 1 °C), indicating that they are predominantly in the austenitic phase at body temperature. Scott et al. reported different results for WO and WOG using DSC [43]. Differences between transformation temperatures determined by BFR and DSC were previously reported [44]. The reasons for these differences are not well understood but it was observed that DSC measures the transformation temperatures in the absence of any external stress, and the transformation temperatures may be affected by residual stress, thermal impedance of the sample geometry and surface condition [45]. Also in DSC, cutting and stacking an acceptable sample weight can affect the stress state and the thermal impedance of the wire in the test cell [45].

WOG showed the highest A_f temperature (25 ± 2 °C) among the tested samples. Alapati et al. reported that M-wire, from which WO instruments are made, shows the three NiTi crystalline phases, martensite, R-phase and austenite [46]. WO showed a wider range between A_s and A_f, indicating the possibility that the phase transformation within the WO instrument is a multi-stage transformation [47]. WOG had an A_f temperature (25 ± 2 °C) higher than that of the used room temperature (20 ± 2 °C) and was expected to outlast WO instruments when operated at room temperature; however, our results showed that this happened only in severely curved canals with a 3 mm radius. In 5 mm canals, no significant difference between WO and WOG was observed in terms of time needed to fracture, which is different from previously reported results [48]. When comparing these two file systems, it should be taken in consideration that they have a different taper, different cross-section designs and are made from different alloys, so these factors can affect their cyclic fatigue resistance [35].

5. Conclusions

Body temperature directly affects the cyclic fatigue resistance of all tested single-file systems and is considered as a factor altering the cyclic fatigue resistance of NiTi endodontic instruments. Single-file systems operated in reciprocation are more resistant to cyclic fatigue failure than those operated in full rotation. The active austenite transformation temperature determined by BFR allows for testing samples in a pre-bent or pre-stressed state, effectively imitating the stresses exerted on endodontic instruments in curved root canals. This emphasizes the ability of BFR to mimic real-world conditions, making it a more relevant and practical method for evaluating the transformation temperature of endodontic instruments in their final product form.

Author Contributions: Conceptualization, E.Y.; Methodology, E.Y.; Software, M.G.; Investigation, E.Y.; Resources, M.G.; Writing—original draft, E.Y.; Writing—review & editing, H.J., S.J. and C.B.; Supervision, H.J. and C.B.; Funding acquisition, C.B. All authors have read and agreed to the published version of the manuscript.

Funding: This research was funded by the BONFOR-Scimed scholarship from the Medical Faculty of the University of Bonn.

Data Availability Statement: Data are contained within the article.

Acknowledgments: The authors would like to thank Anna Weber for her help with SEM imaging.

Conflicts of Interest: The authors declare no conflicts of interest.

Appendix A

Testing block manufacturing: A custom-made block made from stainless steel for medical use (A316) was used for this study, where the artificial canals were engraved using spark erosion processing. The depth of all canals was greater than the shaft diameter of all the files used. All canals were manufactured as described by Plotino et al., 2010 [49], with a relief of 0.1 mm greater than the tested files diameter to ensure the free rotation/reciprocation of the files inside the canals and to exclude fracture of the files due to torsional overload. The metal block had a glass cover fixed with two screws that allowed visualization of the files while testing to record the time needed to fracture and prevented file slippage. The inner surface of the artificial canals had a chrome-hardening layer with a thickness of 20 μm to avoid any dimensional changes in the canals due to friction and wear.

References

1. Walia, H.; Brantley, W.A.; Gerstein, H. An Initial Investigation of the Bending and Torsional Properties of Nitinol Root Canal Files. *J. Endod.* **1988**, *14*, 346–351. [CrossRef]
2. Haapasalo, M.; Shen, Y. Evolution of Nickel–Titanium Instruments: From Past to Future. *Endod. Top.* **2013**, *29*, 3–17. [CrossRef]
3. Shen, Y.; Coil, J.M.; Haapasalo, M. Defects in Nickel-Titanium Instruments after Clinical Use. Part 3: A 4-Year Retrospective Study from an Undergraduate Clinic. *J. Endod.* **2009**, *35*, 193–196. [CrossRef] [PubMed]
4. Sattapan, B.; Nervo, G.; Palamara, J.; Messer, H. Defects in Rotary Nickel-Titanium Files after Clinical Use. *J. Endod.* **2000**, *26*, 161–165. [CrossRef]
5. Knowles, K.I.; Hammond, N.B.; Biggs, S.G.; Ibarrola, J.L. Incidence of Instrument Separation Using LightSpeed Rotary Instruments. *J. Endod.* **2006**, *32*, 14–16. [CrossRef]
6. McGuigan, M.B.; Louca, C.; Duncan, H.F. Endodontic Instrument Fracture: Causes and Prevention. *Br. Dent. J.* **2013**, *214*, 341–348. [CrossRef] [PubMed]
7. De-Deus, G.; Leal Vieira, V.T.; Nogueira da Silva, E.J.; Lopes, H.; Elias, C.N.; Moreira, E.J. Bending Resistance and Dynamic and Static Cyclic Fatigue Life of Reciproc and WaveOne Large Instruments. *J. Endod.* **2014**, *40*, 575–579. [CrossRef]
8. Tsujimoto, M.; Irifune, Y.; Tsujimoto, Y.; Yamada, S.; Watanabe, I.; Hayashi, Y. Comparison of Conventional and New-Generation Nickel-Titanium Files in Regard to Their Physical Properties. *J. Endod.* **2014**, *40*, 1824–1829. [CrossRef]
9. Lopes, H.P.; Gambarra-Soares, T.; Elias, C.N.; Siqueira, J.F.; Inojosa, I.F.J.; Lopes, W.S.P.; Vieira, V.T.L. Comparison of the Mechanical Properties of Rotary Instruments Made of Conventional Nickel-Titanium Wire, M-Wire, or Nickel-Titanium Alloy in R-Phase. *J. Endod.* **2013**, *39*, 516–520. [CrossRef]
10. Azimi, S.; Delvari, P.; Hajarian, H.C.; Saghiri, M.A.; Karamifar, K.; Lotfi, M. Cyclic Fatigue Resistance and Fractographic Analysis of Race and Protaper Rotary NiTi Instruments. *Iran. Endod. J.* **2011**, *6*, 80–86.
11. Dagna, A.; Poggio, C.; Beltrami, R.; Colombo, M.; Chiesa, M.; Bianchi, S. Cyclic Fatigue Resistance of OneShape, Reciproc, and WaveOne: An in Vitro Comparative Study. *J. Conserv. Dent.* **2014**, *17*, 250. [CrossRef] [PubMed]
12. Lin, C.; Wang, Z.; Yang, X.; Zhou, H. Experimental Study on Temperature Effects on NiTi Shape Memory Alloys under Fatigue Loading. *Materials* **2020**, *13*, 573. [CrossRef]
13. Tian, Q.; Wu, J.-S. Characteristic of Specific Heat Capacity of NiTi Alloy Phases. *Trans. Nonferrous Met. Soc. China (Engl. Ed.)* **2000**, *10*, 737–740.
14. Thompson, S.A. An Overview of Nickel–Titanium Alloys Used in Dentistry. *Int. Endod. J.* **2000**, *33*, 297–310. [CrossRef]
15. Pruett, J.P.; Clement, D.J.; Carnes, D.L. Cyclic Fatigue Testing of Nickel-Titanium Endodontic Instruments. *J. Endod.* **1997**, *23*, 77–85. [CrossRef]
16. Pedullà, E.; Rosa, G.R.M.L.; Virgillito, C.; Rapisarda, E.; Kim, H.-C.; Generali, L. Cyclic Fatigue Resistance of Nickel-Titanium Rotary Instruments According to the Angle of File Access and Radius of Root Canal. *J. Endod.* **2020**, *46*, 431–436. [CrossRef]
17. *ASTM F2082-03*; Standard Test Method for Determination of Transformation Temperature of Nickel-Titanium Shape Memory Alloys by Bend and Free Recovery. ASTM International: West Conshohocken, PA, USA, 2017. Available online: https://www.astm.org/f2082-03.html (accessed on 1 December 2023).
18. Yared, G. Canal Preparation Using Only One Ni-Ti Rotary Instrument: Preliminary Observations. *Int. Endod. J.* **2008**, *41*, 339–344. [CrossRef]
19. Grande, N.M.; Ahmed, H.M.A.; Cohen, S.; Bukiet, F.; Plotino, G. Current Assessment of Reciprocation in Endodontic Preparation: A Comprehensive Review—Part I: Historic Perspectives and Current Applications. *J. Endod.* **2015**, *41*, 1778–1783. [CrossRef]
20. Kiefner, P.; Ban, M.; De-Deus, G. Is the Reciprocating Movement per Se Able to Improve the Cyclic Fatigue Resistance of Instruments? *Int. Endod. J.* **2014**, *47*, 430–436. [CrossRef] [PubMed]
21. Pérez-Higueras, J.J.; Arias, A.; de la Macorra, J.C. Cyclic Fatigue Resistance of K3, K3XF, and Twisted File Nickel-Titanium Files under Continuous Rotation or Reciprocating Motion. *J. Endod.* **2013**, *39*, 1585–1588. [CrossRef] [PubMed]
22. Gambarini, G.; Gergi, R.; Naaman, A.; Osta, N.; Al Sudani, D. Cyclic Fatigue Analysis of Twisted File Rotary NiTi Instruments Used in Reciprocating Motion. *Int. Endod. J.* **2012**, *45*, 802–806. [CrossRef]
23. Castelló-Escrivá, R.; Alegre-Domingo, T.; Faus-Matoses, V.; Román-Richon, S.; Faus-Llácer, V.J. In Vitro Comparison of Cyclic Fatigue Resistance of ProTaper, WaveOne, and Twisted Files. *J. Endod.* **2012**, *38*, 1521–1524. [CrossRef]
24. Pedullà, E.; Grande, N.M.; Plotino, G.; Gambarini, G.; Rapisarda, E. Influence of Continuous or Reciprocating Motion on Cyclic Fatigue Resistance of 4 Different Nickel-Titanium Rotary Instruments. *J. Endod.* **2013**, *39*, 258–261. [CrossRef]
25. Varghese, N.O.; Pillai, R.; Sujathen, U.-N.; Sainudeen, S.; Antony, A.; Paul, S. Resistance to Torsional Failure and Cyclic Fatigue Resistance of ProTaper Next, WaveOne, and Mtwo Files in Continuous and Reciprocating Motion: An in Vitro Study. *J. Conserv. Dent.* **2016**, *19*, 225–230. [CrossRef] [PubMed]
26. de Vasconcelos, R.A.; Murphy, S.; Carvalho, C.A.T.; Govindjee, R.G.; Govindjee, S.; Peters, O.A. Evidence for Reduced Fatigue Resistance of Contemporary Rotary Instruments Exposed to Body Temperature. *J. Endod.* **2016**, *42*, 782–787. [CrossRef]
27. Grande, N.M.; Plotino, G.; Silla, E.; Pedullà, E.; DeDeus, G.; Gambarini, G.; Somma, F. Environmental Temperature Drastically Affects Flexural Fatigue Resistance of Nickel-Titanium Rotary Files. *J. Endod.* **2017**, *43*, 1157–1160. [CrossRef] [PubMed]
28. Dosanjh, A.; Paurazas, S.; Askar, M. The Effect of Temperature on Cyclic Fatigue of Nickel-Titanium Rotary Endodontic Instruments. *J. Endod.* **2017**, *43*, 823–826. [CrossRef] [PubMed]

29. Arias, A.; Hejlawy, S.; Murphy, S.; de la Macorra, J.C.; Govindjee, S.; Peters, O.A. Variable Impact by Ambient Temperature on Fatigue Resistance of Heat-Treated Nickel Titanium Instruments. *Clin. Oral Investig.* **2018**, *23*, 1101–1108. [CrossRef] [PubMed]
30. Plotino, G.; Grande, N.M.; Testarelli, L.; Gambarini, G.; Castagnola, R.; Rossetti, A.; Özyürek, T.; Cordaro, M.; Fortunato, L. Cyclic Fatigue of Reciproc and Reciproc Blue Nickel-Titanium Reciprocating Files at Different Environmental Temperatures. *J. Endod.* **2018**, *44*, 1549–1552. [CrossRef] [PubMed]
31. Savitha, S.; Sharma, S.; Kumar, V.; Chawla, A.; Vanamail, P.; Logani, A. Effect of Body Temperature on the Cyclic Fatigue Resistance of the Nickel–Titanium Endodontic Instruments: A Systematic Review and Meta-Analysis of in Vitro Studies. *J. Conserv. Dent.* **2022**, *25*, 338–346. [CrossRef] [PubMed]
32. Estrela, C.; Bueno, M.R.; Sousa-Neto, M.D.; Pécora, J.D. Method for Determination of Root Curvature Radius Using Cone-Beam Computed Tomography Images. *Braz. Dent. J.* **2008**, *19*, 114–118. [CrossRef]
33. Haïkel, Y.; Serfaty, R.; Bateman, G.; Senger, B.; Allemann, C. Dynamic and Cyclic Fatigue of Engine-Driven Rotary Nickel-Titanium Endodontic Instruments. *J. Endod.* **1999**, *25*, 434–440. [CrossRef] [PubMed]
34. Grande, N.M.; Plotino, G.; Pecci, R.; Bedini, R.; Malagnino, V.A.; Somma, F. Cyclic Fatigue Resistance and Three-Dimensional Analysis of Instruments from Two Nickel–Titanium Rotary Systems. *Int. Endod. J.* **2006**, *39*, 755–763. [CrossRef] [PubMed]
35. Parashos, P.; Messer, H.H. Rotary NiTi Instrument Fracture and Its Consequences. *J. Endod.* **2006**, *32*, 1031–1043. [CrossRef] [PubMed]
36. Guillén, R.E.; Nabeshima, C.K.; Caballero-Flores, H.; Cayón, M.R.; Mercadé, M.; Cai, S.; Machado, M.E.D.L.; Guillén, R.E.; Nabeshima, C.K.; Caballero-Flores, H.; et al. Evaluation of the WaveOne Gold and One Shape New Generation in Reducing Enterococcus Faecalis from Root Canal. *Braz. Dent. J.* **2018**, *29*, 249–253. [CrossRef]
37. Klymus, M.E.; Alcalde, M.P.; Vivan, R.R.; Só, M.V.R.; de Vasconselos, B.C.; Duarte, M.A.H. Effect of Temperature on the Cyclic Fatigue Resistance of Thermally Treated Reciprocating Instruments. *Clin. Oral Investig.* **2019**, *23*, 3047–3052. [CrossRef]
38. Jamleh, A.; Alghaihab, A.; Alfadley, A.; Alfawaz, H.; Alqedairi, A.; Alfouzan, K. Cyclic Fatigue and Torsional Failure of EdgeTaper Platinum Endodontic Files at Simulated Body Temperature. *J. Endod.* **2019**, *45*, 611–614. [CrossRef]
39. Keskin, C.; Sivas Yilmaz, Ö.; Keleş, A.; Inan, U. Comparison of Cyclic Fatigue Resistance of Rotate Instrument with Reciprocating and Continuous Rotary Nickel–Titanium Instruments at Body Temperature in Relation to Their Transformation Temperatures. *Clin. Oral Investig.* **2020**, *25*, 151–157. [CrossRef]
40. Kurt, S.M.; Kaval, M.E.; Serefoglu, B.; Demirci, G.K.; Çalışkan, M.K. Cyclic Fatigue Resistance and Energy Dispersive X-Ray Spectroscopy Analysis of Novel Heat-Treated Nickel–Titanium Instruments at Body Temperature. *Microsc. Res. Tech.* **2020**, *83*, 790–794. [CrossRef]
41. Yum, J.-W.; Park, J.-K.; Hur, B.; Kim, H.-C. Comparative Analysis of Various Corrosive Environmental Conditions for NiTi Rotary Files. *J. Korean Acad. Conserv. Dent.* **2008**, *33*, 377–388. [CrossRef]
42. Pelton, A.R.; Dicello, J.; Miyazaki, S. Optimisation of Processing and Properties of Medical Grade Nitinol Wire. *Minim. Invasive Ther. Allied Technol.* **2000**, *9*, 107–118. [CrossRef]
43. Scott, R.; Arias, A.; Macorra, J.C.; Govindjee, S.; Peters, O.A. Resistance to Cyclic Fatigue of Reciprocating Instruments Determined at Body Temperature and Phase Transformation Analysis. *Aust. Endod. J.* **2019**, *45*, 400–406. [CrossRef]
44. Zhang, S.; Denton, M.; Desai, P.; Licht, G.; Fariabi, S. Phase Transition Sequence and Af Determination in Nickel–Titanium Shape Memory Alloys. In *SMST-2004: Proceedings of the International Conference on Shape Memory and Superelastic Technologies, Baden-Baden, Germany, 3–7 October 2004*; ASM International: Almere, The Netherlands, 2006; pp. 21–27, ISBN 978-1-61503-118-4.
45. Cadelli, A.; Manjeri, R.M.; Sczerzenie, F.; Coda, A. Uniaxial Pre-Strain and Free Recovery (UPFR) as a Flexible Technique for Nitinol Characterization. *Shap. Mem. Superelasticity* **2016**, *2*, 86–94. [CrossRef]
46. Alapati, S.B.; Brantley, W.A.; Iijima, M.; Clark, W.A.T.; Kovarik, L.; Buie, C.; Liu, J.; Johnson, W.B. Metallurgical Characterization of a New Nickel-Titanium Wire for Rotary Endodontic Instruments. *J. Endod.* **2009**, *35*, 1589–1593. [CrossRef]
47. Carroll, M.C.; Somsen, C.; Eggeler, G. Multiple-Step Martensitic Transformations in Ni-Rich NiTi Shape Memory Alloys. *Scr. Mater.* **2004**, *50*, 187–192. [CrossRef]
48. Özyürek, T. Cyclic Fatigue Resistance of Reciproc, WaveOne, and WaveOne Gold Nickel-Titanium Instruments. *J. Endod.* **2016**, *42*, 1536–1539. [CrossRef]
49. Plotino, G.; Grande, N.M.; Mazza, C.; Petrovic, R.; Testarelli, L.; Gambarini, G. Influence of Size and Taper of Artificial Canals on the Trajectory of NiTi Rotary Instruments in Cyclic Fatigue Studies. *Oral Surg. Oral Med. Oral Pathol. Oral Radiol. Endodontology* **2010**, *109*, e60–e66. [CrossRef]

Disclaimer/Publisher's Note: The statements, opinions and data contained in all publications are solely those of the individual author(s) and contributor(s) and not of MDPI and/or the editor(s). MDPI and/or the editor(s) disclaim responsibility for any injury to people or property resulting from any ideas, methods, instructions or products referred to in the content.

Article

Mechanical Properties and Root Canal Shaping Ability of a Nickel–Titanium Rotary System for Minimally Invasive Endodontic Treatment: A Comparative In Vitro Study

Hayate Unno, Arata Ebihara *, Keiko Hirano, Yuka Kasuga, Satoshi Omori, Taro Nakatsukasa, Shunsuke Kimura, Keiichiro Maki and Takashi Okiji

Department of Pulp Biology and Endodontics, Division of Oral Health Sciences,
Graduate School of Medical and Dental Sciences, Tokyo Medical and Dental University (TMDU), 1-5-45 Yushima, Bunkyo-ku, Tokyo 113-8549, Japan
* Correspondence: a.ebihara.endo@tmd.ac.jp; Tel.: +81-3-5803-5494

Abstract: Selection of an appropriate nickel–titanium (NiTi) rotary system is important for minimally invasive endodontic treatment, which aims to preserve as much root canal dentin as possible. This study aimed to evaluate selected mechanical properties and the root canal shaping ability of TruNatomy (TRN), a NiTi rotary system designed for minimally invasive endodontic shaping, in comparison with existing instruments: HyFlex EDM (HEDM), ProTaper Next (PTN), and WaveOne Gold (WOG). Load values measured with a cantilever bending test were ranked as TRN < HEDM < WOG < PTN ($p < 0.05$). A dynamic cyclic fatigue test revealed that the number of cycles to fracture was ranked as HEDM > WOG > TRN > PTN ($p < 0.05$). Torque and vertical force generated during instrumentation of J-shaped artificial resin canals were measured using an automated instrumentation device connected to a torque and vertical force measuring system; TRN exhibited smaller torque and vertical force values in most comparisons with the other instruments. The canal centering ratio for TRN was smaller than or comparable to that for the other instruments except for WOG at the apex level. Under the present experimental conditions, TRN showed higher flexibility and lower torque and vertical force values than the other instruments.

Keywords: austenite; fatigue fractures; martensite; nickel–titanium; root canal preparation; torque

1. Introduction

Nickel–titanium (NiTi) rotary instruments have gained popularity for root canal instrumentation because of their high flexibility [1] and ability to maintain root canal curvature [2,3]. In recent years, various technical advances in metallurgy, geometry, and kinematics have enabled the development of different NiTi rotary systems with improved efficiency and predictable safety [4]. In particular, heat treatment has been widely applied to NiTi alloys to improve the mechanical characteristics of NiTi rotary instruments, including flexibility and resistance to cyclic fatigue, by adjusting phase transition temperatures to induce the growth of the ductile martensite phase and R-phase [5]. For these newly introduced NiTi rotary instruments, scientific evidence regarding preclinical parameters, such as root canal shaping ability and fracture resistance for their appropriate clinical application, should be gathered.

ProTaper Next (PTN; Dentsply Sirona, Ballaigues, Switzerland) is a well-investigated heat-treated NiTi rotary system made of M-wire with improved cyclic fatigue resistance [6]. PTN has an off-centered square cross-section with a variable taper [7]. HyFlex EDM (HEDM; Coltène-Whaledent, Altstätten, Switzerland) is manufactured from distinctively heat-treated CM-wire, together with electrical discharge machining [8], and shows improved cutting efficiency and cyclic fatigue resistance [9]. The cross-section of HEDM is variable: rectangular, trapezoidal and triangular in the apical, middle, and coronal portions, respectively. WaveOne Gold (WOG; Dentsply Sirona, Ballaigues, Switzerland) is a representative NiTi reciprocating system manufactured from Gold-wire with a parallelogram cross-sectional shape with two cutting edges. WOG reportedly exhibits higher flexibility and cyclic fatigue resistance [10], and a comparable ability to maintain canal curvature [11] as its predecessor, WaveOne (Dentsply Sirona).

The concept of minimally invasive endodontic treatment involves preserving as much of the root dentin as possible [12]. The use of instruments with a small diameter and taper is advocated to limit the amount of dentin cutting and preserve the cervical dentin [13] to improve resistance to tooth fracture [14]. TruNatomy (TRN; Dentsply Sirona) is a recently launched NiTi rotary system with this concept incorporated into its design features. TRN is produced from a proprietary post-manufacture heat-treated NiTi alloy with a regressive taper, maximum flute diameter of 0.8 mm, and a parallelogram, off-centered cross-sectional design [15–17]. The TRN system consists of an orifice shaping file, a glide path file, and three files for root canal shaping with a single-length technique. TRN is reported to be more resistant to cyclic fatigue in comparison with instruments made of differently heat-treated alloys, such as PTN [17] and Vortex Blue (Dentsply Sirona) [16]. The degree of apical canal deviation induced by TRN is smaller than [18] or similar to [19] several other NiTi rotary systems.

More information is, however, required to determine what advantages TRN offers over other NiTi rotary and reciprocating systems for performing minimally invasive endodontic shaping. Therefore, with the aim of evaluating the properties of TRN, this study examined this instrument's flexibility, resistance to cyclic fatigue, generation of torque and vertical force during instrumentation, and root canal shaping ability, in comparison with those of other contemporary NiTi rotary and reciprocating systems. The null hypothesis was that there are no significant differences in bending load, dynamic cyclic fatigue resistance, torque and vertical force values during canal instrumentation, and root canal centering ability among the tested instruments.

2. Materials and Methods

Ethical approval is not applicable to this study because it does not contain any studies with human or animal subjects/materials.

2.1. Sample Size Estimation

G*Power software (version 3.1.9.7; Heinrich Heine Universität, Düsseldorf, Germany) was used to determine the sample size required. A priori analysis of variance (ANOVA; fixed effects, omnibus, one-way) was selected from the F-test family. The effect size was set at 0.6 based on data from a previous study [20]. An alpha-type error of 0.05 and a power beta of 0.85 were also specified. The required sample size of 10 per group was obtained.

2.2. Bending Test

TRN Prime (#26/0.04 taper, Dentsply Sirona), HEDM One File (#25/0.08-0.04 taper, Coltène-Whaledent), PTN X2 (#25/0.06-0.07 taper, Dentsply Sirona), and WOG Primary (#25/0.07-0.03 taper, Dentsply Sirona) were tested, using a self-made cantilever bending tester described in previous studies [21,22] (n = 10 for each model) (Figure 1). The file was fixed at 7 mm from the tip, and loaded at a position 2 mm from the tip at a speed of 1.0 mm/min until there was 3 mm of displacement. The bending load was measured at 0.5 mm and 2.0 mm of displacement. All experiments were performed at room temperature.

2.3. Dynamic Cyclic Fatigue Test

The same instruments as in the bending test were evaluated. A self-made cyclic fatigue tester with a movable test stand (MH2-500N; IMADA, Aichi, Japan) and the X-Smart Plus endodontic motor (Dentsply Sirona) were used [23] (Figure 2). An artificial root canal made of stainless steel and designed with a 1.5 mm diameter, a 60° curvature, 3 mm radius of curvature, and the center of the curvature at 5 mm from the tip of the instrument was used [24]. The instruments were rotated as per the manufacturers' recommendations (500 rpm, 1.5 N·cm in TRN; 400 rpm, 2.4 N·cm in HEDM; 300 rpm, 2.0 N·cm in PTN; and the WaveOne Gold setting in WOG), while moving the handpiece with an axial up-and-down motion of 2 mm amplitude at 300 mm/min. The canal was lubricated with silicon oil (KF-96-100CS, Shin-Etsu Chemical, Tokyo, Japan). The length of time to fracture was measured, and the number of cycles to failure (NCF) was determined as the number of revolutions (rpm) × time to fracture (seconds). All experiments were performed at room temperature.

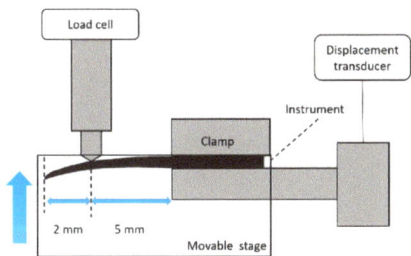

Figure 1. Schematic drawing showing the cantilever bending tester used in the experiment.

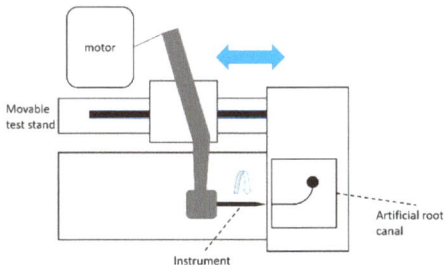

Figure 2. Schematic drawing showing the cyclic fatigue tester used in the experiment.

2.4. Measurement of Torque and Vertical Force

A self-made automated root canal instrumentation device connected to a torque and vertical force measuring system was employed as described in previous studies [20,25–27] (Figure 3). Briefly, this device consisted of the X-Smart Plus motor and a test stand (MX2-500N, IMADA, Aichi, Japan), and the torque and vertical force measuring system was configured with a built-in load cell (LUX-B-ID, Kyowa, Tokyo, Japan) for determining vertical force, and strain gauges (KFG-2-120-D31-11, Kyowa, Tokyo, Japan) for determining torque. The movable device of the stand, to which the handpiece was attached, was programmed to make an up-and-down motion at a speed of 50 mm/min for 2 s downward and 1 s upward [25,28].

J-shaped artificial resin root canals (0.02 taper, 0.1 mm apical foramen diameter, 45° curvature, 16 mm full working length; Endo Training Bloc, Dentsply Sirona; n = 40) were randomly assigned to Groups TRN, HEDM, PTN and WOG (n = 10, each) and instrumented according to the sequence recommended for narrow canals by the manufacturers. After coronal flaring to a depth of 12 mm, the canal was subjected to automated instrumentation consisting of glide path preparation to the full working length (16 mm), followed by a two-instrument full-length instrumentation where the working lengths were set incrementally at 14, 15, and 16 mm for the two instruments. After each instrumentation, a #10 K-file (Zipperer, Munich, Germany) was used to verify the patency, and canal irrigation was performed with 1 mL of distilled water. The canal was lubricated with RC-Prep (Premier Dental, Plymouth Meeting, PA, USA). Torque (cutting direction) and vertical force (downward and upward) were monitored during the automated root canal instrumentation and the maximum values were statistically analyzed.

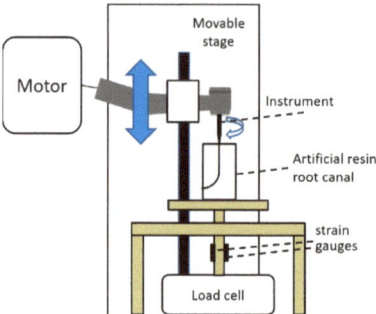

Figure 3. Schematic drawing of the automated root canal instrumentation device and the torque and vertical force measuring system used in the experiment.

The rotary instruments used were as follows: TRN Orifice Modifier (#20/0.08 taper), TRN Glider (#17/0.02 taper), TRN Small (#20/0.04 taper) and TRN Prime (#26/0.04 taper) in Group TRN; HEDM Orifice Opener (#25/0.12 taper), HEDM Glide Path File (#10/0.05 taper), HEDM Preparation File (#20/0.05 tape) and HEDM One File (#25/0.08–0.04 taper) in Group HEDM; ProTaper SX (#19/0.04 taper; Dentsply Sirona), ProGlider (#16/0.02 taper; Dentsply Sirona), PTN X1 (#17/0.04 taper) and PTN X2 (#25/0.06 taper) in Group PTN; and ProTaper Gold SX (#19/0.04 taper), WOG Glider (#15/0.06 taper), WOG Small (#20/0.07 taper), and WOG Prime (#25/0.07 taper) in Group WOG. The rotational speeds and torque-limit values were set as per the manufacturers' recommendations as follows: 500 rpm, 1.5 N·cm for TRN; 300 rpm, 1.8 N·cm for HEDM Glide Path File; 400 rpm, 2.4 N·cm for HEDM; 250 rpm, 4.0 N·cm for ProTaper SX; 300 rpm, 2.0 N·cm for PTN; 300 rpm, 3.0 N·cm for ProTaper Gold SX; and the WOG mode for WOG. All experiments were performed at room temperature.

2.5. Root Canal Centering Ability

Preoperative (before the glide path preparation) and postoperative digital images of the artificial resin canals described above were obtained using a digital microscope (VH-8000, Keyence, Osaka, Japan). The images were then overlaid and analyzed with image analysis software (Adobe Photoshop Elements 2021, San Jose, CA, USA) [28]. The canal centering ratio was calculated at 0, 0.5, 1, 2, and 3 mm from the apex using the following formula:

$$(X-Y)/Z$$

where:
X = amount of removed resin material from the outer canal wall
Y = amount of removed resin material from the inner canal wall
Z = root canal diameter after instrumentation [29].

The post-instrumentation canal deviation becomes smaller as the centering ratio approaches 0. All instrumentation was performed at room temperature.

2.6. Statistical Analysis

SPSS software (version 27.0; IBM, Armonk, NY, USA) was used to determine statistical differences at a significance level of 5%. Data normality and homogeneous variance were verified using the Shapiro–Wilk test and the Levene's F test, respectively. The values obtained in the bending test, torque and vertical force values, and the canal centering ratio were analyzed with two-way ANOVA and the Tukey post-hoc test. The NCF values were analyzed by one-way ANOVA and the Tukey post-hoc test.

3. Results

3.1. Bending Loads

At a 0.5 mm deflection, PTN withstood a significantly larger load than the other instruments ($p < 0.05$; Figure 4). No significant difference was observed among TRN, HEDM, and WOG ($p > 0.05$). At a 2.0 mm deflection, the load values were ranked as TRN < HEDM < WOG < PTN ($p < 0.05$; Figure 4).

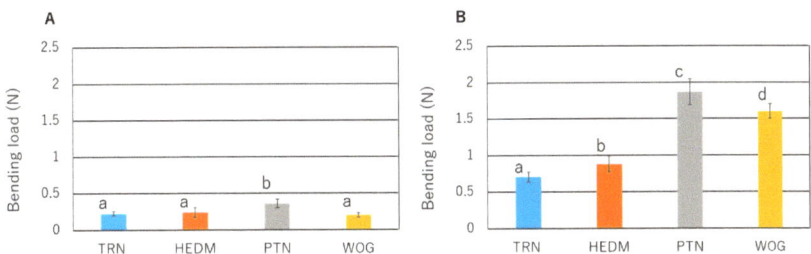

Figure 4. Bending load values (N) of the tested NiTi rotary instruments at a deflection of 0.5 mm (elastic region: **A**) and 2.0 mm (superelastic region: **B**). Values are means and standard deviations (n = 10). Groups with different letters in each testing condition are significantly different ($p < 0.05$; one-way ANOVA and Tukey test).

3.2. Dynamic Cyclic Fatigue Resistance

The NCF value was the highest in HEDM, followed by WOG, TRN, and PTN ($p < 0.05$; Figure 5).

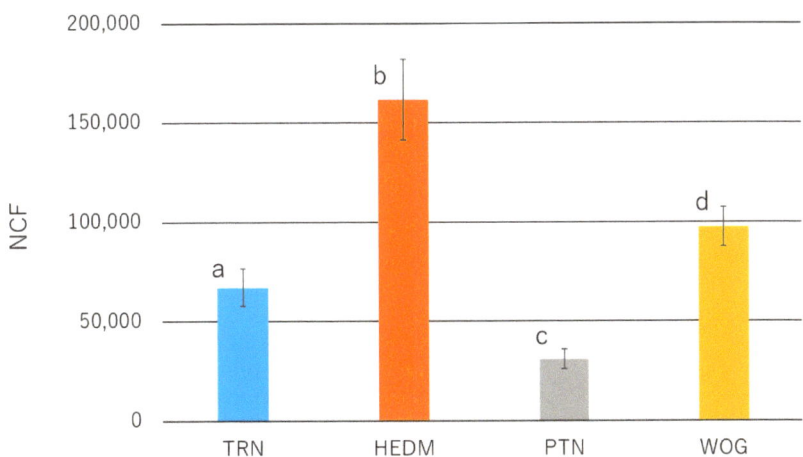

Figure 5. Number of cycles to failure (NCF) of NiTi rotary instruments subjected to a dynamic cyclic fatigue test. Values are means and standard deviations (n = 10). Groups with different letters are significantly different ($p < 0.05$; one-way ANOVA and Tukey test).

3.3. Torque and Vertical Force

Figure 6 shows the maximum torque and vertical force values produced by each instrument in each group. No file fracture occurred during root canal instrumentation.

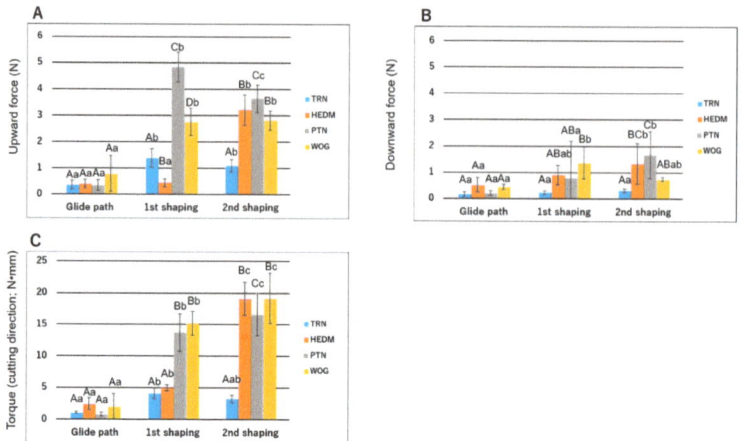

Figure 6. Maximum upward force (**A**), downward force (**B**), and torque in the cutting direction (**C**) generated during root canal instrumentation. Values are means and standard deviations (n = 10). Different uppercase letters in each instrumentation sequence in each panel indicate that the values are significantly different in intergroup comparisons ($p < 0.05$; two-way ANOVA and Tukey test). Different lowercase letters in each panel indicate that the values are significantly different within the same instrument ($p < 0.05$; two-way ANOVA and Tukey test).

Upward force values for the first shaping instruments were ranked as HEDM < TRN < WOG < PTN ($p < 0.05$). Among the second shaping instruments, TRN showed a significantly lower upward force compared with the other instruments ($p < 0.05$). Intragroup comparisons revealed that all shaping instruments, except the first instrument in HEDM, showed significantly higher values than the corresponding glide path instrument.

Regarding the downward force, WOG demonstrated significantly higher values than TRN ($p < 0.05$) when the first shaping instruments were compared. Among the second shaping instruments, PTN and HEDM exhibited significantly higher values than the other instruments ($p < 0.05$). Intragroup comparisons revealed that the first shaping instrument in WOG and the second shaping instrument in HEDM and PTN exhibited significantly higher values than the corresponding glide path instruments ($p < 0.05$), while the three instruments in Group TRN showed similar values.

Torque values in the cutting direction among the first shaping instruments were ranked as PTN and WOG > TRN and HEDM ($p < 0.05$). Torque values among the second shaping instruments were TRN < PTN < HEDM and WOG. Within Group HEDM, Group PTN, and Group WOG, torque values were ranked as glide path instrument < first shaping instrument < second shaping instrument ($p < 0.05$). Within Group TRN, the first shaping instrument generated a significantly higher torque value than the glide path instrument ($p < 0.05$).

3.4. Canal Centering Ratio

At the 0 mm level, Group WOG showed a significantly smaller value than the other groups ($p < 0.05$; Figure 7). At the 0.5 mm level, significantly higher and lower values were found in Group WOG and Group PTN, respectively ($p < 0.05$; Figure 7). At the 1 mm and 2 mm levels, the ratios of Group PTN were significantly higher than those of the other groups ($p < 0.05$; Figure 7). No significant differences were found among the groups at 3 mm from the apex.

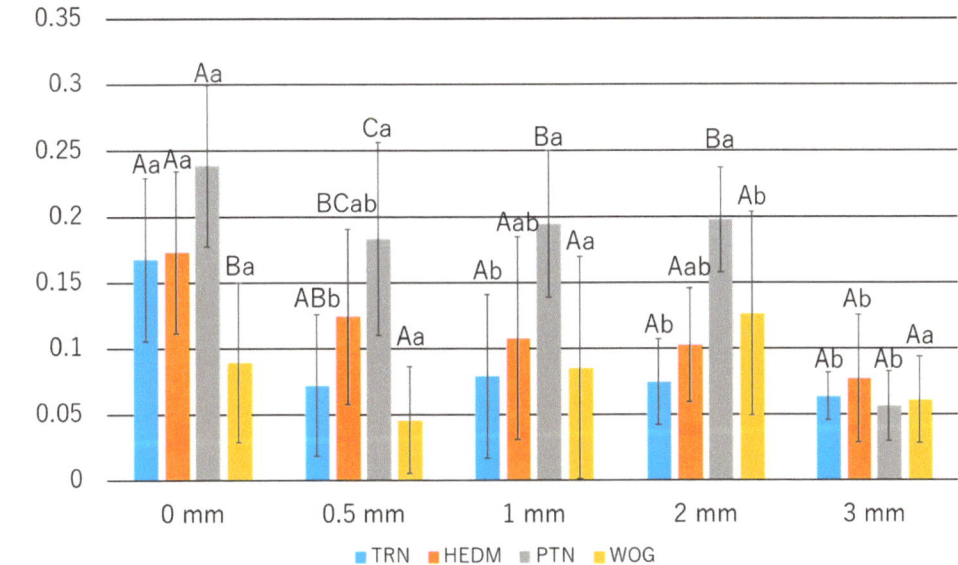

Figure 7. Canal centering ratios. Values are means and standard deviations (n = 10). Different uppercase letters in the measurement level indicate that the values are significantly different in intergroup comparisons ($p < 0.05$; two-way ANOVA and Tukey test). Different lowercase letters indicate that the values are significantly different within the same instrument ($p < 0.05$; two-way ANOVA and Tukey test).

4. Discussion

The aim of minimally invasive endodontic treatment is to minimize the amount of dentin cutting, particularly near the cervical area [17,19]. However, restricted coronal enlargement in a curved canal may make an instrument more stressed and the apical portion of a canal more prone to deviation during instrumentation [19,30]. Thus, selection of appropriate instruments with sufficient fracture resistance and canal centering ability is important to avoid iatrogenic events.

In this study, a series of experiments testing flexibility, resistance to cyclic fatigue, torque and vertical force generation, and root canal shaping ability were conducted in an attempt to obtain a comprehensive understanding of the characteristics of TRN as a NiTi rotary system suitable for minimally invasive endodontic treatment. The results demonstrated significant differences in all the tests across the tested instruments; in particular, TRN exhibited a significantly smaller bending load, and significantly lower torque and vertical force values in several comparisons with the other instruments. Thus, the null hypothesis was rejected.

Various factors such as geometry [31], heat treatment [1,32], and the alloy manufacturing process [5,33] affect the mechanical properties of NiTi rotary instruments. Because no existing NiTi rotary systems are identical to TRN in terms of overall design, control instruments were selected from dimensionally similar, well-studied systems, as in several other studies [17,19,34–36]. This made the comparisons sufficiently meaningful to test the performance of TRN in comparison with instruments with different metallurgy and rotational modes.

In the bending test, the displacement at 0.5 mm corresponds to the elastic region, and at 2.0 mm to the superelastic region [37]. In this study, TRN exhibited the smallest bending load in the superelastic region, which corroborates the high flexibility of TRN as reported in a previous study [38]. Several factors influence the flexibility of NiTi rotary instruments, including geometry (e.g., size, cross-sectional design, core diameter, and pitch length) [39,40] and metallurgy [5,33]. Regarding metallurgy, differential scanning calorimetric studies have reported that the austenite starting and finishing temperatures of TRN are 11.8 °C and 29.2 °C, respectively [38], in contrast with those reported for PTN (0.2 °C and 51.4 °C; [41]), HEDM (45.47 °C and 51.55 °C; [42]), and WOG (8.5 °C and 51.6 °C; [41]). Thus, at the tested temperature, TRN was in the mixed phase of austenite, R-phase, and martensite, and HEDM was in the highest martensite composition, which indicates that the lowest flexibility of TRN cannot be explained by its phase composition. Thus, the highest flexibility of TRN may be primarily attributable to the geometric factors of TRN, such as the smaller wire diameter (0.8 mm in TRN vs. 1.2 mm in the other instruments) [17], the parallelogram cross-sectional design, which may contribute to decreasing the core diameter [40], and the smallest taper in the tip region.

The dynamic cyclic fatigue test incorporates axial movement mimicking a pecking motion and thus better simulates clinical conditions compared to the static cyclic fatigue test [43–46]. In the dynamic test, the fatigue life of NiTi instruments is longer than that in the static test because the axial movement releases the stress concentrated at the curvature [43,44,47]. In this study, NCF values were ranked as PTN < TRN < WOG < HEDM, which can be attributed to differences in several factors, including heat treatment, kinematic movement, and cross-sectional design [48]. The NCF value for TRN was significantly higher than that for PTN, which is consistent with a previous study [17]. TRN and PTN differ in heat treatment (post- vs. pre-manufacturing treatment), taper, and cross-sectional design (parallelogram vs. rectangle), while it is unclear which of these factors contribute more to the higher NCF of TRN than PTN. Of interest was that the NCF of TRN was lower than that of HEDM and WOG, which is contradictory to the notion that highly flexible instruments yield higher cyclic fatigue resistance. Additionally, the present results do not agree with the findings that instruments with a smaller diameter perform better in the cyclic fatigue test [49]. These contradictory results may be explained by the fact that the martensite-rich metallurgy in HEDM [48,50] and the reciprocal motion in WOG [51] counteracted the impact of the diameter and flexibility, resulting in the higher cyclic fatigue life of these instruments.

NiTi rotary instruments exert torque and downward force to cut dentin, and if the torsional stress accumulated in the instruments exceeds the elastic limit, torsional fracture may occur [52]. The upward force represents the screw-in force, which may expose a rotating file to a risk of sudden torsional stress leading to separation [29,31,53]. In this study, the torque and force produced by all the glide path instruments were low. Among the shaping instruments, TRN in general produced lower torque and vertical force than the other instruments. These findings are consistent with a previous finding that TRN produced smaller torque and vertical force than PTN [17], and may be attributed to the cumulative effect of various influencing factors. Smaller taper and diameter [54] and higher flexibility [31,55] are among detrimental factors associated with lower torque/force production. The off-centered parallelogram cross-sectional design of TRN, which contacts with the root canal wall at one or two points during instrumentation, may also play a vital role in its lower screw-in force generation [53]. In addition, pre-instrumentation canal size/taper relative to the dimension of an instrument may have an impact on torque/force generation. This may explain the significant increase, compared with the previous instrument, of torque and/or upward force in the first shaping instrument in TRN, PTN, and WOG and the second shaping instrument in HEDM and PTN; these instruments are greater in both size and taper compared to the previous instrument. These finding imply which instrument(s) are stressed, and thus may clinically require careful manipulation during the instrumentation sequence of each system.

The canal centering ratio demonstrated for TRN was, in general, smaller than or comparable to that for the other instruments at each measuring point, which may be explained by the fact that flexible instruments cause less canal deviation [56,57]. At the 0 mm point, however, WOG showed a significantly smaller ratio than TRN. This may be due to the reciprocating motion of WOG, which disengages a bound instrument during rotation in the non-cutting direction and thus reduces the contact time of the blades to the canal wall, leading to improved canal centering ability [58].

Collectively, the present findings suggest that TRN has several mechanical properties that favor its use under the concept of minimally invasive endodontic treatment, such as higher flexibility and lower torque/force values than the other instruments. However, this study is not without limitations. One limitation is the use of artificial resin canals to evaluate torque and vertical force generation and shaping ability; although resin canals offer a highly standardized root canal morphology, their hardness is lower and their surface smoother than those of real canals and thus less force may be required for instrumentation [59]. Another potential limitation is the test temperature (room temperature), under which the metallurgical microstructure and mechanical properties of NiTi instruments may be different from those under clinical conditions [38]. Accurate temperature measurement of NiTi instruments during root canal shaping under clinical conditions has not yet been reported; in future experiments, temperature taken in real time should be taken into account. Further study is required to determine how NiTi instruments perform differently under different temperature settings between the laboratory and clinical conditions. Incorporation of tests for other relevant parameters, such as torsional resistance, surface hardness, and cutting efficiency, may facilitate more comprehensive analysis. Future studies should also be carried out to evaluate the influence of the root canal anatomy, which is highly variable [60] with several factors, including degree of curvature, that may influence cyclic fatigue resistance [24] and torque and force generation [52] of NiTi rotary instruments. Increasing knowledge of the characteristics of NiTi rotary instruments will be beneficial for clinicians in making better selection of these instruments according to individual cases.

5. Conclusions

This study examined the flexibility, resistance to cyclic fatigue, generation of torque and vertical force during instrumentation, and root canal shaping ability of TRN, in comparison with those of HEDM, PTN, and WOG. The following conclusions were drawn:

- The bending load values were not significantly different between TRN, HEDM, and WOG at a 0.5 mm deflection. At a 2.0 mm deflection, the load values were ranked as TRN < HEDM < WOG < PTN.
- The number of cycles to fracture was ranked as HEDM > WOG > TRN > PTN.
- During instrumentation of J-shaped artificial resin canals, TRN showed smaller torque and vertical force values in most comparisons with the other instruments.
- The canal centering ratio for TRN was smaller than or comparable to that for the other instruments, except for WOG at the apex level.
- TRN showed higher flexibility and lower torque/force values than the other instruments, which may favor its clinical use under the concept of minimally invasive endodontic treatment.

Author Contributions: Conceptualization, H.U., A.E. and T.O.; methodology, H.U. and A.E.; software, H.U.; validation, H.U.; formal analysis, H.U.; investigation, H.U., K.H., Y.K., S.O., T.N., S.K. and K.M.; resources, H.U.; data curation, H.U., A.E., K.H., Y.K., S.O., T.N., S.K., K.M. and T.O.; writing—original draft preparation, H.U.; writing—review and editing, A.E. and T.O.; visualization, H.U.; supervision, A.E. and T.O.; project administration, A.E.; funding acquisition, H.U. All authors have read and agreed to the published version of the manuscript.

Funding: This research was funded by a contract research grant (grant no. 11BA100267) from Tokyo Medical and Dental University.

Institutional Review Board Statement: Not applicable.

Informed Consent Statement: Not applicable.

Data Availability Statement: The data presented in this study are available on request from the corresponding author.

Acknowledgments: We thank Helen Jeays, BDSc AE, for editing a draft of this manuscript.

Conflicts of Interest: The authors declare no conflict of interest.

References

1. Ebihara, A.; Yahata, Y.; Miyara, K.; Nakano, K.; Hayashi, Y.; Suda, H. Heat treatment of nickel-titanium rotary endodontic instruments: Effects on bending properties and shaping abilities. *Int. Endod. J.* **2011**, *44*, 843–849. [CrossRef] [PubMed]
2. Gambill, J.M.; Alder, M.; Del Rio, C.E. Comparison of nickel-titanium and stainless steel hand-file instrumentation using computed tomography. *J. Endod.* **1996**, *22*, 369–375. [CrossRef]
3. Peters, O.A. Current challenges and concepts in the preparation of root canal systems: A review. *J. Endod.* **2004**, *30*, 559–567. [CrossRef] [PubMed]
4. Arias, A.; Peters, O.A. Present status and future directions: Canal shaping. *Int. Endod. J.* **2022**, *55*, 637–655. [CrossRef]
5. Shen, Y.; Zhou, H.M.; Zheng, Y.F.; Peng, B.; Haapasalo, M. Current challenges and concepts of the thermomechanical treatment of nickel-titanium instruments. *J. Endod.* **2013**, *39*, 163–172. [CrossRef]
6. Pereira, E.S.J.; Gomes, R.O.; Leroy, A.M.F.; Singh, R.; Peters, O.A.; Bahia, M.G.A.; Buono, V.T.L. Mechanical behavior of M-Wire and conventional NiTi wire used to manufacture rotary endodontic instruments. *Dent. Mater.* **2013**, *29*, e318–e324. [CrossRef]
7. Koçak, M.M.; Çiçek, E.; Koçak, S.; Sağlam, B.C.; Furuncuoğlu, F. Comparison of ProTaper Next and HyFlex instruments on apical debris extrusion in curved canals. *Int. Endod. J.* **2016**, *49*, 996–1000. [CrossRef]
8. Iacono, F.; Pirani, C.; Generali, L.; Bolelli, G.; Sassatelli, P.; Lusvarghi, L.; Gandolfi, M.G.; Giorgini, L.; Prati, C. Structural analysis of HyFlex EDM instruments. *Int. Endod. J.* **2017**, *50*, 303–313. [CrossRef]
9. Pedullà, E.; Lo Savio, F.; Boninelli, S.; Plotino, G.; Grande, N.M.; La Rosa, G.; Rapisarda, E. Torsional and cyclic fatigue resistance of a new nickel-titanium instrument manufactured by electrical discharge machining. *J. Endod.* **2016**, *42*, 156–159. [CrossRef]
10. Fangli, T.; Maki, K.; Kimura, S.; Nishijo, M.; Tokita, D.; Ebihara, A.; Okiji, T. Assessment of mechanical properties of WaveOne Gold Primary reciprocating instruments. *Dent. Mater. J.* **2019**, *38*, 490–495. [CrossRef]
11. Bürklein, S.; Fluch, S.; Schäfer, E. Shaping ability of reciprocating single-file systems in severely curved canals: WaveOne and Reciproc versus WaveOne Gold and Reciproc blue. *Odontology* **2019**, *107*, 96–102. [CrossRef] [PubMed]
12. Gluskin, A.H.; Peters, C.I.; Peters, O.A. Minimally invasive endodontics: Challenging prevailing paradigms. *Br. Dent. J.* **2014**, *216*, 347–353. [CrossRef] [PubMed]
13. Rundquist, B.D.; Versluis, A. How does canal taper affect root stresses? *Int. Endod. J.* **2006**, *39*, 226–237. [CrossRef] [PubMed]
14. Tang, W.; Wu, Y.; Smales, R.J. Identifying and reducing risks for potential fractures in endodontically treated teeth. *J. Endod.* **2010**, *36*, 609–617. [CrossRef]
15. Bürklein, S.; Zupanc, L.; Donnermeyer, D.; Tegtmeyer, K.; Schäfer, E. Effect of core mass and alloy on cyclic fatigue resistance of different nickel-titanium endodontic instruments in matching artificial canals. *Materials* **2021**, *14*, 5734. [CrossRef]
16. Elnaghy, A.M.; Elsaka, S.E.; Mandorah, A.O. In vitro comparison of cyclic fatigue resistance of TruNatomy in single and double curvature canals compared with different nickel-titanium rotary instruments. *BMC Oral Health* **2020**, *20*, 38. [CrossRef]
17. Peters, O.A.; Arias, A.; Choi, A. Mechanical properties of a novel nickel-titanium root canal instrument: Stationary and dynamic tests. *J. Endod.* **2020**, *46*, 994–1001. [CrossRef]
18. Kim, H.; Jeon, S.J.; Seo, M.S. Comparison of the canal transportation of ProTaper GOLD, WaveOne GOLD, and TruNatomy in simulated double-curved canals. *BMC Oral Health* **2021**, *21*, 533. [CrossRef]
19. Kabil, E.; Katić, M.; Anić, I.; Bago, I. Micro-computed evaluation of canal transportation and centering ability of 5 rotary and reciprocating systems with different metallurgical properties and surface treatments in curved root canals. *J. Endod.* **2021**, *47*, 477–484. [CrossRef]
20. Kyaw, M.S.; Ebihara, A.; Maki, K.; Kimura, S.; Nakatsukasa, T.; Htun, P.H.; Thu, M.; Omori, S.; Okiji, T. Effect of kinematics on the torque/force generation, surface characteristics, and shaping ability of a nickel-titanium rotary glide path instrument: An ex vivo study. *Int. Endod. J.* **2022**, *55*, 531–543. [CrossRef]
21. Miyara, K.; Yahata, Y.; Hayashi, Y.; Tsutsumi, Y.; Ebihara, A.; Hanawa, T.; Suda, H. The influence of heat treatment on the mechanical properties of Ni-Ti file materials. *Dent. Mater. J.* **2014**, *33*, 27–31. [CrossRef] [PubMed]
22. Fukumori, Y.; Nishijyo, M.; Tokita, D.; Miyara, K.; Ebihara, A.; Okiji, T. Comparative analysis of mechanical properties of differently tapered nickel-titanium endodontic rotary instruments. *Dent. Mater. J.* **2018**, *37*, 667–674. [CrossRef] [PubMed]
23. Nakatsukasa, T.; Ebihara, A.; Kimura, S.; Maki, K.; Nisijo, M.; Tokita, D.; Okiji, T. Comparative evaluation of mechanical properties and shaping performance of heat-treated nickel titanium rotary instruments used in the single-length technique. *Dent. Mater. J.* **2021**, *40*, 743–749. [CrossRef] [PubMed]
24. Pruett, J.P.; Clement, D.J.; Carnes, D.L. Cyclic fatigue testing of nickel-titanium endodontic instruments. *J. Endod.* **1997**, *23*, 77–85. [CrossRef]
25. Tokita, D.; Ebihara, A.; Nishijo, M.; Miyara, K.; Okiji, T. Dynamic torque and vertical force analysis during nickel-titanium rotary root canal preparation with different modes of reciprocal rotation. *J. Endod.* **2017**, *43*, 1706–1710. [CrossRef]
26. Kimura, S.; Ebihara, A.; Maki, K.; Nishijo, M.; Tokita, D.; Okiji, T. Effect of optimum torque reverse motion on torque and force generation during root canal instrumentation with crown-down and single-length techniques. *J. Endod.* **2020**, *46*, 232–237. [CrossRef]

27. Htun, P.H.; Ebihara, A.; Maki, K.; Kimura, S.; Nishijo, M.; Tokita, D.; Okiji, T. Comparison of torque, force generation and canal shaping ability between manual and nickel-titanium glide path instruments in rotary and optimum glide path motion. *Odontology* **2020**, *108*, 188–193. [CrossRef]
28. Maki, K.; Ebihara, A.; Kimura, S.; Nishijo, M.; Tokita, D.; Okiji, T. Effect of different speeds of up-and-down motion on canal centering ability and vertical force and torque generation of nickel-titanium rotary instruments. *J. Endod.* **2019**, *45*, 68–72. [CrossRef]
29. Maki, K.; Ebihara, A.; Kimura, S.; Nishijo, M.; Tokita, D.; Miyara, K.; Okiji, T. Enhanced root canal-centering ability and reduced screw-in force generation of reciprocating nickel-titanium instruments with a post-machining thermal treatment. *Dent. Mater. J.* **2020**, *39*, 251–255. [CrossRef]
30. Vorster, M.; Van der Vyver, P.J.; Markou, G. The effect of different molar access cavity designs on root canal shaping times using rotation and reciprocation instruments in mandibular first molars. *J. Endod.* **2022**, *48*, 887–892. [CrossRef]
31. Ha, J.H.; Cheung, G.S.P.; Versluis, A.; Lee, C.J.; Kwak, S.W.; Kim, H.C. 'Screw-in' tendency of rotary nickel-titanium files due to design geometry. *Int. Endod. J.* **2015**, *48*, 666–672. [CrossRef] [PubMed]
32. Miyai, K.; Ebihara, A.; Hayashi, Y.; Doi, H.; Suda, H.; Yoneyama, T. Influence of phase transformation on the torsional and bending properties of nickel-titanium rotary endodontic instruments. *Int. Endod. J.* **2006**, *39*, 119–126. [CrossRef] [PubMed]
33. Zupanc, J.; Vahdat-Pajouh, N.; Schäfer, E. New thermomechanically treated NiTi alloys—A review. *Int. Endod. J.* **2018**, *51*, 1088–1103. [CrossRef]
34. Riyahi, A.M.; Bashiri, A.; Alshahrani, K.; Alshahrani, S.; Alamri, H.M.; Al-Sudani, D. Cyclic fatigue comparison of TruNatomy, Twisted File, and ProTaper Next rotary systems. *Int. J. Dent.* **2020**, *2020*, 3190938. [CrossRef] [PubMed]
35. Mustafa, R.; Al Omari, T.; Al-Nasrawi, S.; Al Fodeh, R.; Dkmak, A.; Haider, J. Evaluating in vitro performance of novel nickel-titanium rotary system (TruNatomy) based on debris extrusion and preparation time from severely curved canals. *J. Endod.* **2021**, *47*, 976–981. [CrossRef] [PubMed]
36. Reddy, B.N.; Murugesan, S.; Basheer, S.N.; Kumar, R.; Kumar, V.; Selvaraj, S. Comparison of cyclic fatigue resistance of novel TruNatomy files with conventional endodontic files: An in vitro SEM study. *J. Contemp. Dent. Pract.* **2021**, *22*, 1243–1249.
37. Yahata, Y.; Yoneyama, T.; Hayashi, Y.; Ebihara, A.; Doi, H.; Hanawa, T.; Suda, H. Effect of heat treatment on transformation temperatures and bending properties of nickel-titanium endodontic instruments. *Int. Endod. J.* **2009**, *42*, 621–626. [CrossRef]
38. Silva, E.; Martins, J.N.R.; Ajuz, N.C.; Antunes, H.S.; Vieira, V.T.L.; Braz Fernandes, F.M.; Belladonna, F.G.; Versiani, M.A. A multimethod assessment of a new customized heat-treated nickel-titanium rotary file system. *Materials* **2022**, *15*, 5288. [CrossRef]
39. Viana, A.C.D.; De Melo, M.C.C.; Bahia, M.G.D.; Buono, V.T.L. Relationship between flexibility and physical, chemical, and geometric characteristics of rotary nickel-titanium instruments. *Oral Surg. Oral Med. Oral Pathol. Oral Radiol. Endod.* **2010**, *110*, 527–533. [CrossRef]
40. Versluis, A.; Kim, H.C.; Lee, W.; Kim, B.M.; Lee, C.J. Flexural stiffness and stresses in nickel-titanium rotary files for various pitch and cross-sectional geometries. *J. Endod.* **2012**, *38*, 1399–1403. [CrossRef]
41. Hou, X.M.; Yang, Y.J.; Qian, J. Phase transformation behaviors and mechanical properties of NiTi endodontic files after gold heat treatment and blue heat treatment. *J. Oral Sci.* **2020**, *63*, 8–13. [CrossRef] [PubMed]
42. Arias, A.; Macorra, J.C.; Govindjee, S.; Peters, O.A. Correlation between temperature-dependent fatigue resistance and differential scanning calorimetry analysis for 2 contemporary rotary instruments. *J. Endod.* **2018**, *44*, 630–634. [CrossRef] [PubMed]
43. Lopes, H.P.; Elias, C.N.; Vieira, M.V.; Siqueira, J.F., Jr.; Mangelli, M.; Lopes, W.S.; Vieira, V.T.L.; Alves, F.R.F.; Oliveira, J.C.M.; Soares, T.G. Fatigue life of Reciproc and Mtwo instruments subjected to static and dynamic tests. *J. Endod.* **2013**, *39*, 693–696. [CrossRef] [PubMed]
44. Keleş, A.; Eymirli, A.; Uyanik, O.; Nagas, E. Influence of static and dynamic cyclic fatigue tests on the lifespan of four reciprocating systems at different temperatures. *Int. Endod. J.* **2019**, *52*, 880–886. [CrossRef] [PubMed]
45. Haïkel, Y.; Serfaty, R.; Bateman, G.; Senger, B.; Allemann, C. Dynamic and cyclic fatigue of engine-driven rotary nickel-titanium endodontic instruments. *J. Endod.* **1999**, *25*, 434–440. [CrossRef]
46. Yao, J.H.; Schwartz, S.A.; Beeson, T.J. Cyclic fatigue of three types of rotary nickel-titanium files in a dynamic model. *J. Endod.* **2006**, *32*, 55–57. [CrossRef]
47. Thu, M.; Ebihara, A.; Maki, K.; Miki, N.; Okiji, T. Cyclic fatigue resistance of rotary and reciprocating nickel-titanium instruments subjected to static and dynamic tests. *J. Endod.* **2020**, *46*, 1752–1757. [CrossRef]
48. Chi, D.L.; Zhang, Y.J.; Lin, X.W.; Tong, Z.C. Cyclic fatigue resistance for six types of nickel titanium instruments at artificial canals with different angles and radii of curvature. *Dent. Mater. J.* **2021**, *40*, 1129–1135. [CrossRef]
49. Faus-Llácer, V.; Kharrat, N.H.; Ruiz-Sanchez, C.; Faus-Matoses, I.; Zubizarreta-Macho, A.; Faus-Matoses, V. The effect of taper and apical diameter on the cyclic fatigue resistance of rotary endodontic files using an experimental electronic device. *Appl. Sci.* **2021**, *11*, 863. [CrossRef]
50. Pirani, C.; Iacono, F.; Generali, L.; Sassatelli, P.; Nucci, C.; Lusvarghi, L.; Gandolfi, M.G.; Prati, C. HyFlex EDM: Superficial features, metallurgical analysis and fatigue resistance of innovative electro discharge machined NiTi rotary instruments. *Int. Endod. J.* **2016**, *49*, 483–493. [CrossRef]
51. De-Deus, G.; Moreira, E.J.L.; Lopes, H.P.; Elias, C.N. Extended cyclic fatigue life of F2 ProTaper instruments used in reciprocating movement. *Int. Endod. J.* **2010**, *43*, 1063–1068. [CrossRef] [PubMed]

52. Thu, M.; Ebihara, A.; Adel, S.; Okiji, T. Analysis of torque and force induced by rotary nickel-titanium instruments during root canal preparation: A systematic review. *Appl. Sci.* **2021**, *11*, 3079. [CrossRef]
53. Kwak, S.W.; Lee, C.J.; Kim, S.K.; Kim, H.C.; Ha, J.H. Comparison of screw-in forces during movement of endodontic files with different geometries, alloys, and kinetics. *Materials* **2019**, *12*, 1506. [CrossRef] [PubMed]
54. Sattapan, B.; Palamara, J.E.A.; Messer, H.H. Torque during canal instrumentation using rotary nickel-titanium files. *J. Endod.* **2000**, *26*, 156–160. [CrossRef]
55. Kwak, S.W.; Ha, J.H.; Lee, C.J.; El Abed, R.; Abutahun, I.H.; Kim, H.C. Effects of pitch length and heat treatment on the mechanical properties of the glide path preparation instruments. *J. Endod.* **2016**, *42*, 788–792. [CrossRef]
56. Pinheiro, S.R.; Alcalde, M.P.; Vivacqua-Gomes, N.; Bramante, C.M.; Vivan, R.R.; Duarte, M.A.H.; Vasconcelos, B.C. Evaluation of apical transportation and centring ability of five thermally treated NiTi rotary systems. *Int. Endod. J.* **2018**, *51*, 705–713. [CrossRef]
57. Hieawy, A.; Haapasalo, M.; Zhou, H.; Wang, Z.J.; Shen, Y. Phase transformation behavior and resistance to bending and cyclic fatigue of ProTaper Gold and ProTaper Universal instruments. *J. Endod.* **2015**, *41*, 1134–1138. [CrossRef]
58. Elashiry, M.M.; Saber, S.E.; Elashry, S.H. Comparison of shaping ability of different single-file systems using microcomputed tomography. *Eur. J. Dent.* **2020**, *14*, 70–76. [CrossRef]
59. Lim, K.C.; Webber, J. The validity of simulated root canals for the investigation of the prepared root canal shape. *Int. Endod. J.* **1985**, *18*, 240–246. [CrossRef]
60. Mirza, M.B.; Gufran, K.; Alhabib, O.; Alafraa, O.; Alzahrani, F.; Abuelqomsan, M.S.; Karobari, M.I.; Alnajei, A.; Afroz, M.M.; Akram, S.M.; et al. CBCT based study to analyze and classify root canal morphology of maxillary molars—A retrospective study. *Eur. Rev. Med. Pharmacol. Sci.* **2022**, *26*, 6550–6560.

Article

The Washout Resistance of Bioactive Root-End Filling Materials

Joanna Falkowska [1,*], Tomasz Chady [2], Włodzimierz Dura [1], Agnieszka Droździk [3], Małgorzata Tomasik [3], Ewa Marek [1], Krzysztof Safranow [4] and Mariusz Lipski [1,*]

1. Department of Preclinical Conservative Dentistry and Preclinical Endodontics, Pomeranian Medical University in Szczecin, Al. Powstańców Wlkp. 72, 70-111 Szczecin, Poland; wlodzimierz.dura@pum.edu.pl (W.D.); ewa.marek@pum.edu.pl (E.M.)
2. Faculty of Electrical Engineering, West Pomeranian University of Technology in Szczecin, Sikorsky 37 St., 70-313 Szczecin, Poland; tchady@zut.edu.pl
3. Department of Interdisciplinary Dentistry, Pomeranian Medical University in Szczecin, Al. Powstańców Wlkp. 72, 70-111 Szczecin, Poland; agnieszka.drozdzik@pum.edu.pl (A.D.); malgorzata.tomasik@pum.edu.pl (M.T.)
4. Department of Biochemistry and Medical Chemistry, Pomeranian Medical University, Al. Powstańców Wlkp. 72, 70-111 Szczecin, Poland; krzysztof.safranow@pum.edu.pl
* Correspondence: joanna.falk@vp.pl (J.F.); mariusz.lipski@pum.edu.pl (M.L.)

Citation: Falkowska, J.; Chady, T.; Dura, W.; Droździk, A.; Tomasik, M.; Marek, E.; Safranow, K.; Lipski, M. The Washout Resistance of Bioactive Root-End Filling Materials. *Materials* 2023, 16, 5757. https://doi.org/10.3390/ma16175757

Academic Editor: Luigi Generali

Received: 12 July 2023
Revised: 18 August 2023
Accepted: 19 August 2023
Published: 23 August 2023

Copyright: © 2023 by the authors. Licensee MDPI, Basel, Switzerland. This article is an open access article distributed under the terms and conditions of the Creative Commons Attribution (CC BY) license (https://creativecommons.org/licenses/by/4.0/).

Abstract: Fast-setting bioactive cements were developed for the convenience of retrograde fillings during endodontic microsurgery. This in vitro study aimed to investigate the effect of irrigation on the washout of relatively fast-setting materials (Biodentine, EndoCem Zr, and MTA HP) in comparison with MTA Angelus White and IRM in an apicectomy model. Washout resistance was assessed using artificial root ends. A total of 150 samples (30 for each material) were tested. All samples were photographed using a microscope, and half of them were also scanned. The samples were irrigated and immersed in saline for 15 min. Then the models were evaluated. Rinsing and immersing the samples immediately after root-end filling and after 3 min did not disintegrate the fillings made of all tested materials except Biodentine. Root-end fillings made of Biodentine suffered significant damage both when rinsing was performed immediately and 3 min after the filling. Quantitative assessment of washed material resulted in a slight loss of IRM, EndoCem MTA Zr, and MTA HP. MTA Angelus White showed a slightly greater washout. Rinsing and immersion of Biodentine restorations resulted in their significant destruction. Under the conditions of the current study, the evaluated materials, excluding Biodentine, showed good or relatively good washout resistance.

Keywords: bioactive cements; washout resistance; retrograde root canal filling

1. Introduction

Surgical endodontic treatment is performed when orthograde root canal therapy is unsuccessful and retreatment is either impossible or useless [1]. Endodontic surgery usually involves exposure of the apex, the removal of pathological periapical tissue, root-end resection (apicoectomy), root-end cavity preparation, and the placement of a retrograde filling material [1]. The success rate of this surgical procedure is above 90% and is close to that of orthograde root canal treatment [2–4].

Several materials have been used in periapical surgery, including amalgam, intermediate restorative material (IRM), super ethoxybenzoic acid (Super-EBA), glass ionomer cements, polycarboxylate cements, zinc phosphate cements, calcium phosphate cements, composite resins, and calcium silicate cements, but none of them have the characteristics of an ideal root-end filling material [5–9]. An ideal root-end filling material should provide a long-term hermetic seal resulting from resistance to washout and dissolution in periapical fluids and good adaptation to the walls of retrograde preparation. It must be non-irritating, non-toxic, non-carcinogenic, and biocompatible. For clinical applications, it should be easy to manipulate and radiopaque [1,5,7,8].

Mineral trioxide aggregate (MTA) is usually used as the material of choice for retrograde root canal filling because it has most of the properties of an ideal root-end filling material, such as bioactivity and biocompatibility, antimicrobial effect, good sealing ability, and setting in a humid environment [1,7–15]. Many authors also claim that these materials evoke a positive tissue response to promote the regeneration of the periodontium [16–18]. However, MTA has some drawbacks, such as a long setting time, difficult application, low resistance to compression and flexion, and a high cost [6,19–22]. Many studies have also reported that MTA leads to tooth discoloration [23–25].

In order to improve the clinical results of the treatment, many new calcium silicate materials have been developed. Biodentine® (Septodont, Saint Maur des Fossés, France), a new dental substitute, was introduced on the market in 2009. Biodentine is composed of a powder component (mainly calcium silicates) and a liquid component (water with calcium chloride and a water-soluble polymer). The powder is placed in a capsule, while the liquid is in an ampoule. Mixing is achieved using a triturator for 30 s at 4000–4200 rpm. According to the manufacturer, the initial setting time is about 12–15 min [26]. However, some authors [27] estimated the final setting time of this material to be 85 min. The consistencies of Biodentine and phosphoric cement are similar, which makes Biodentine easy to apply [28]. As for MTA, the indications and clinical applications for Biodentine are indirect/direct pulp capping, pulpotomy, apexogenesis, apexification (apical plug), root and crown perforation, resorption repair, and retrograde root-end filling [28–30]. Because Biodentine was released at the end of 2009, many laboratory studies have been published so far with this material. The literature is still, however, inconclusive concerning the superiority of Biodentine over MTA as a root-end filling material in apical surgery. In some reports, Biodentine produced a better seal than MTA, whereas other studies failed to confirm any superiority of Biodentine or found it to be inferior to MTA when filling the retrograde cavities. A recently published meta-analysis showed that there is a lack of scientific evidence for the superiority of Biodentine over MTA as a retrograde filling material in apical surgery [31].

Both MTA (e.g., ProRoot MTA, Dentsply Sirona, Charlotte, NC, USA, MTA Angelus, Angelus, Londrina, Brazil) and Biodentine are calcium silicate cements. In the process of cement setting, hydration is a very important reaction. The main product of this reaction is calcium silicate hydrate. Some authors conduct research on the effect of the presence of aluminum on the effectiveness of this reaction. Although the presence of aluminum may affect the efficiency of the reaction, it may also decrease its biocompatibility and lengthen the setting time. Among tested materials, aluminum is present in MTA-type materials, but fully synthetic Biodentine does not contain it [31,32].

Another newly developed type of bioactive material for end filling is EndoCem Zr (Maruchi, Wonju, Republic of Korea). It has been introduced as an MTA-derived pozzolan cement (a naturally occurring siliceous and aluminous material of volcanic origin) [33]. EndoCem Zr has a short setting time of 4 min and favorable manipulation properties [33,34]. Pozzolan, when mixed with water, undergoes a reaction with calcium hydroxide to form calcium silicate hydrate, similar to that produced by the hydration of MTA. EndoCem/EndoCem Zr sets fast despite no accelerator. The faster setting of the material is probably due to the small particles of cement and thus a larger contact surface with water [35,36]. It has low tooth tissue discoloration potential because it contains zirconium oxide instead of conventional bismuth oxide [34]. However, an in vitro study showed that Endocem Zr was more cytotoxic and associated with lower expression of VEGF and ANG in comparison with mineral trioxide aggregate [37], which was confirmed by histopathologic analysis in a canine model of pulpotomy that showed fewer odontoblast layer formation and fewer calcific barrier formation with greater inflammatory response in comparison with the mineral trioxide aggregate [38].

A new type of MTA, MTA repair high plasticity (MTA HP, Angelus, Londrina, Brazil), has been introduced recently. It is based on the standard MTA formula but contains calcium tungstate ($CaWO_4$) as a radiopacifier and a liquid consisting of water and a plasticizing agent [39,40]. This new composition retains the chemical properties of the original MTA but amends its physical properties (plasticity), making it easier to manipulate and insert into the retrograde cavity than traditional MTA [40,41]. MTA HP compared to ProRoot MTA sets faster (12 min) [39]. The more rapid setting of this calcium silicate cement is explained in the literature by the large surface area of the powder particles and the absence of sulphate phases [42].

One of the characteristics of the ideal material for a retrograde root canal filling is resistance to washout, i.e., the resistance of freshly prepared cement to disintegration upon early contact with fluid [43]. The term washout originated in engineering science and is used to describe washing out the material from freshly mixed cement with water [44]. In dentistry, the washout may have a considerable impact on the result of the treatment because the loss of the filling material may be the reason for microleakage. That is one of the reasons why washout resistance has already been studied by other researchers over the years [45,46]. After filling the root-end preparation, it is recommended that the resection cavity be rinsed gently; bleeding that occurs during periapical surgery is also responsible for the disintegration of the filling [47]. Therefore, the cements used for the retrograde filling should be resistant to washout. This resistance increases with time, although it should be borne in mind that it is challenging to guarantee dryness during surgery for a few minutes.

This in vitro study aimed to investigate the effect of irrigation on washout of relatively fast-setting materials (Biodentine, EndoCem Zr, and MTA HP) in comparison with root-end filling materials that have been on the market for several years, MTA Angelus White, and IRM, in an apicectomy model.

2. Materials and Methods

The following materials were used in this study:

1. Intermediate Restorative Material (IRM; Dentsply Sirona, Charlotte, NC, USA);
2. MTA Angelus White (Angelus, Londrina, Brazil);
3. Biodentine (Septodont, Saint Maur-des-Fossés, Cedex, France);
4. EndoCem Zr (Maruchi, Wonju, Republic of Korea);
5. MTA HP (Angelus, Londrina, Brazil).

Their compositions are outlined in Table 1.

Table 1. Composition of the commercial materials.

Material	Manufacturer	Ingredient	Mixing
IRM	Dentsply Sirona, Charlotte, NC, USA	powder: zinc oxide, poly-methyl methacrylate (PMMA) powder, pigment liquid: eugenol, acetic acid	1 spoon of powder + 1 drop of distilled water (mixed manually on glass slab using a metal spatula, 30 s)
MTA Angelus White	Angelus, Londrina, Brazil	powder: tricalcium silicate, dicalcium silicate, tricalcium aluminate, ferroaluminate tricalcium, calcium oxide, bismuth oxide liquid: distilled water	2 level scoops of powder + 3 drops of liquid (mixed manually on glass slab using a metal spatula, 30 s)
Biodentine	Septodont, Saint-Maur-des-Fossés Cedex, France	powder: tricalcium silicate, dicalcium silicate, calcium carbonate and oxide filler, iron oxide shade, and zirconium oxide liquid: calcium chloride as an accelerator, hydrosoluble polymer water-reducing agent, water	0.7 g capsule of powder + 5 drops of liquid (mixed in the triturator; 30 s; 4000–4200 rpm)
EndoCem Zr	Maruchi, Wonju, Republic of Korea	powder: calcium oxide, silicon dioxide, aluminum oxide, magnesium oxide, ferrous oxide, zirconium oxide liquid: distilled water	0.3 g of powder + 0.12 mL (mixed manually on glass slab using a metal spatula; 30 s)
MTA HP	Angelus, Londrina, Brazil	powder: tricalcium silicate, dicalcium silicate, tricalcium aluminate, calcium oxide, and calcium tungstate liquid: water and plasticizer	0.085 g capsule of powder + 2 drops of liquid (mixed manually on a glass slab using a metal spatula; 30 s)

IRM, MTA Angelus White, EndoCem Zr, and MTA HP powder were mixed manually with their liquid according to recommendations from the manufacturer. Biodentine was mixed in the triturator (Silver MIX, GC Dental, Tokyo, Japan) for 30 s.

2.1. Assessment of Washout Resistance

Washout resistance was assessed using artificial root-end preparations of 1.2 mm × 3 mm (diameter and depth corresponding to the diameter and height of the cylinder) in plastic blocks. For this purpose, the crowns of incisors placed in the model of the mandible (Frasaco, Tettnang, Germany) were prepared to resemble resected roots (Figure 1A), and the bottom of the bony crypts were simulated with the use of silicone impression material (Figure 1B). In total, 150 samples (30 for each material) of freshly prepared materials were placed in artificial root-ends prepared in plastic blocks (Figure 1C). The application time for each material was 30 s. Materials were placed in the plastic blocks by using the MTA+ Applicator (Cerkamed, Stalowa Wola, Poland; https://cerkamed.pl/produkt/aplikator-mta/ accessed on 20 August 2023) and condensed with retro filing plugger No. 1 (Medesy SRL, Maniago, Italy; www.medesy.it/en/products/endodontic-instrument-with-plugger/ accessed on 20 August 2023).

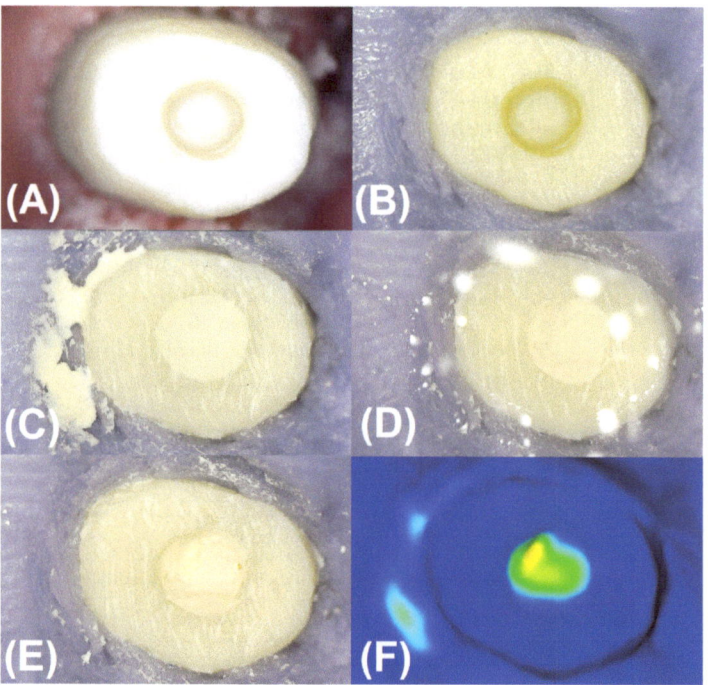

Figure 1. (**A**) The crown of the mandibular incisor (plastic block) after preparation to resemble the resected root with a root-end preparation of 1.2 mm × 3 mm (diameter and depth). (**B**) The bottom of the bony crypts were simulated with the use of silicone impression material. (**C**) Root-end preparation filled with retrograde material. (**D**) Rinsing of the resected root surface and the artificial bony crypt. (**E**). Partial rinsing of the material from the root-end preparation. (**F**) The color model in the form of a depth map obtained as a result of imposing scans.

2.1.1. Experiment 1

Seventy-five samples (15 of each material) were photographed using a Levenhuk DTX 90 microscope (Levenhuk, Inc., Tampa, FL, USA) at 60× magnification. The resolution of the microscope, experimentally determined using a microscopic slide with a micrometer scale, was 10 μm. In addition, several fillings were evaluated by watching the material-wall interface using microscopes with a resolution better than 10 μm (Digital Microscope VHX-7000, Keyence, Osaka, Japan, and Digital Micro Hardness Tester, Model: MHVD—1000IS, INNOVATEST, Wiry, Poland). It was found that the width of the narrowest gaps recorded with the use of the microscope used in the present study ranged from 7–8 to 20–25 μm. If the quality of the filling was questionable, the test sample was replaced with a new one. The photography of the sample (samples were prepared one by one and photographed one by one) was taken immediately after the material application was finished (the time of taking the photography was max. 60 s) and the rinsing started immediately. For this purpose, 5 mL saline (ambient temperature) was applied for 15 s using a disposable syringe and needle of 0.8 mm diameter and 12 mm length. The saline from the needle opening impinged on the edge as in the clinical setting, i.e., the saline flowed over the resected surface, washing over the test material but not directly spraying into it (Figure 1D). After the simulation of the rinsing of the bony crypt, the models were immersed in warm saline (34 °C) for 15 min (simulation of blood flooding the crypt). After this time, the models were removed, gently dried without disrupting the fillings, and then photographed again under a microscope (Figure 1E).

2.1.2. Experiment 2

The remaining 75 samples (15 of each material) were used in this experiment. Apart from photographing the samples under a microscope, the surface of the fillings was additionally scanned using KaVo ARCTICA AutoScan—the 3D dental scanner (KaVo, Biberach, Germany). Scanning and photography lasted 3 min, which can be assumed to be the time of protection against humidity in a clinical setting. Then, the samples were flushed in the same way as in experiment 1 and immersed in warm saline for 15 min, after which time the models were removed, photographed under the microscope, and scanned again using a 3D dental scanner (KaVo ARCTICA AutoScan, KaVo, Biberach, Germany). Then, the recorded scans were superimposed to obtain a color model in the form of a depth map (Figure 1F).

2.2. Qualitative Analysis of the Marginal Adaptation of the Materials to the Walls of the Root-End Preparations and Disintegration of Fillings

The adaptation of the retrograde materials to the walls of root-end preparations and the disintegration of root-end fillings were evaluated based on photographs. A total of 150 coded photographs from samples in groups 1–10 were evaluated by three evaluators who had been calibrated before the assessment. Each evaluator gave an independent score without reference to the other evaluators.

The qualitative analysis of the marginal adaptation of retrograde material to the root-end cavities and washed-out area was based on the following grading criteria (Figure 2):

Score 1: Close marginal approximation of the filling material to the wall of preparation, no gaps present at the material-wall interface, no washed-out area (hollow area) on the material's surface.

Score 2: No gaps present at the material-wall interface, washed-out area (hollow area) on the material's surface.

Score 3: The presence of gaps at the material-wall interface and the washed out area (hollow area) on the material's surface.

Those criteria were created based on the sample evaluation by Tran et al. [48].

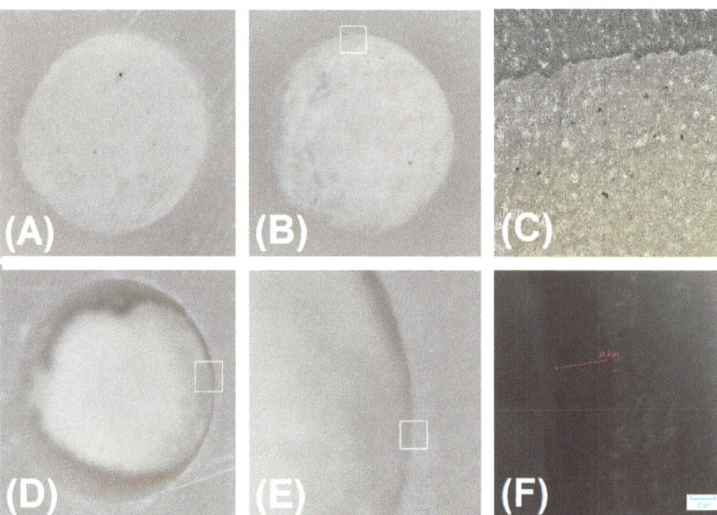

Figure 2. Criteria for marginal adaptation of retrograde material to the root-end cavities and washed-out area: (**A**) Score 1: Close marginal approximation of the filling material to the wall of preparation, no gaps present at the material-wall interface, no washed-out area (hollow area) on the surface of the material; (**B**) Score 2: No gaps present at the material-wall interface, washed-out areas (hollow areas) on the surface of the material. (**C**). View of the rectangular in (**B**)—close marginal approximation of the filling material to the wall of preparation (photography taken at 500× using Digital Microscope VHX-7000, Keyence, Osaka, Japan), (**D**): Score 3: Presence of narrow gaps at the material-wall interface, washed-out area (hollow area) on the surface of the material. (**E**) Extensive washout of the material. (**F**) View of the rectangular in (**E**)—the gap between the material and the wall preparation is 21 μm wide (photography taken at 500× using Digital Micro Hardness Tester, Model: MHVD—1000IS, INNOVATEST, Wiry, Poland).

2.3. Quantitative Assessment of the Volume of Washed out Material

The software written by the authors in the programming language and numeric computing environment Matlab (The MathWorks, Inc., Torrance, CA, USA) was used for the quantitative assessment of the volume of washed-out material. The washed-out material was quantified by superimposing the scans recorded before and after rinsing. In this way, a geometric solid was created in the form of a depth map (a two-dimensional table with the depths of the cavity at a given point). The geometric solid was divided into layers (within a single solid, the layers had the same thickness, e.g., 5.1282 μm; layer thickness within individual solids ranged from 4 to 7 μm) and the layers into cuboids (within a single solid, the width and depth of cuboids were always the same, e.g., 4.486×4.486 μm^2; within individual solids, the width and depth of cuboids ranged from 4 to 7 μm, the height of cuboids depended on the thickness of the layer, e.g., 5.1282 μm) of equal volume. Then the number of cuboids was counted, and the result was multiplied by the volume of the cuboid. In this way, the approximate volume of the solid was determined. The volume is given in mm^3 and as a % of the filling (a cylinder with a diameter of 1.2 mm and a height of 3 mm). In addition, the average value in the depth map table was calculated, which is the average depth of the hollow (Figure 3).

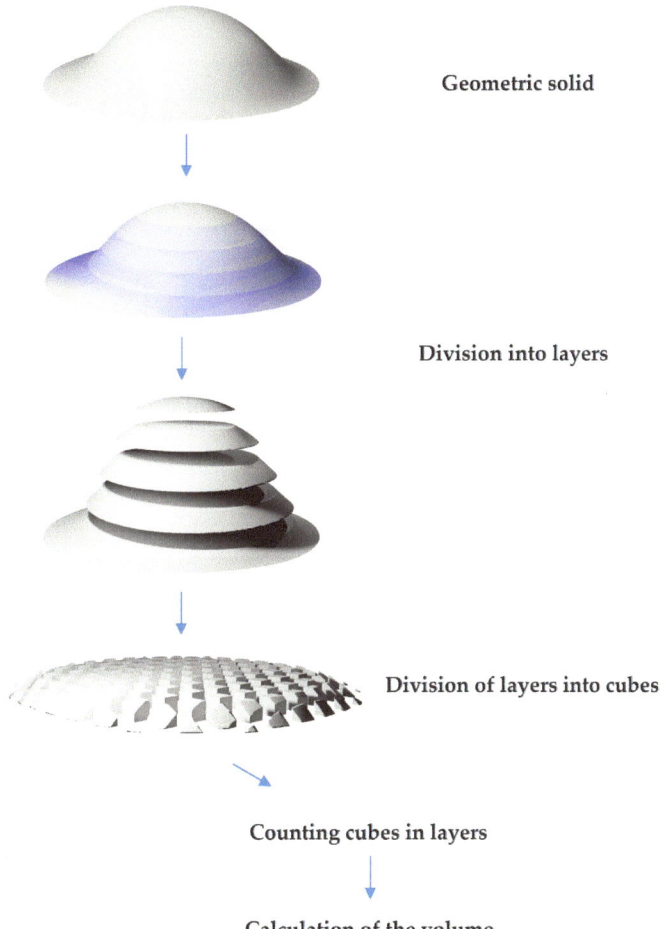

Figure 3. Schematic drawing of quantitative estimation of the volume of the geometric solid.

2.4. Statistical Analysis

The mean score of qualitative analyses by three evaluators was compared between materials or between experiments with a non-parametric Kruskal–Wallis test or Mann–Whitney test. A weighted kappa was used as a measure of the evaluators' assessment concordance. Quantitative washout values were presented as means with a standard deviation (SD). They were compared between materials with ANOVA followed by Tukey's post hoc test after log transformation to obtain the homoscedasticity ($p > 0.05$, Levene test) and normality of distributions ($p > 0.1$, Shapiro–Wilk test) necessary for parametric analysis. A result was considered statistically significant at $p < 0.05$. Statistica 13 was used for statistical analyses.

3. Results

3.1. Qualitative Analysis of the Marginal Adaptation of the Materials to the Walls of the-End Preparations and Disintegration of Fillings

Rinsing and immersion of the samples in the liquid simulated blood immediately after root-end filling did not disintegrate the fillings made of IRM, MTA Angelus White, EndoCem MTA Zr, and MTA HP (Table 2). No disintegration of the fillings made of these materials occurred when the rinsing procedure was performed 3 min after the application of the materials into the root-end cavities (Figures 2–5). On the other hand, root-end fillings made of Biodentine suffered significant damage both when rinsing was performed immediately after the application of materials into the root-end cavity and 3 min after filling (Table 2, Figure 6). There was no significant difference in the degree of damage depending on the time from application to rinsing. The weighted Cohen's kappa statistic showed very good agreement between the three evaluators (Table 3).

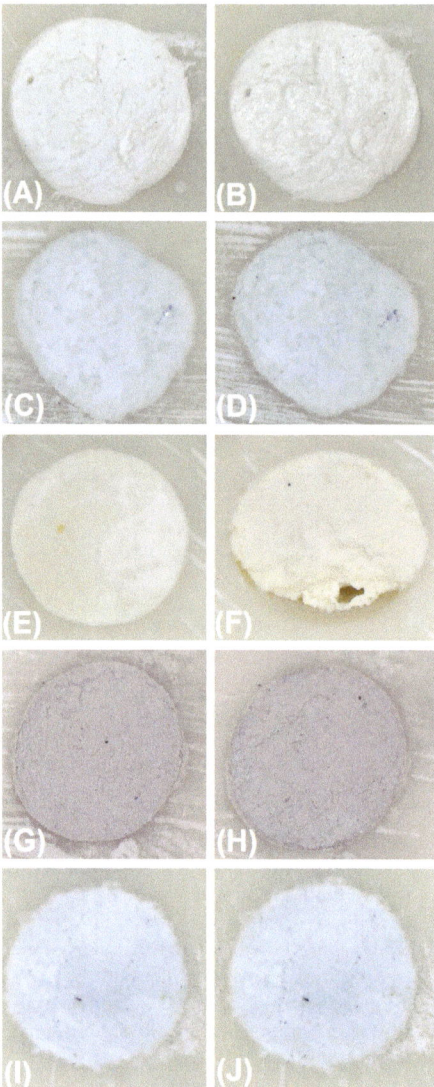

Figure 4. Adaption of retrograde materials to the walls of root-end preparations and disintegration of root-end filling. (**A**)—IRM before and (**B**) after rinsing and immersion of the samples in simulated blood; (**C**) MTA Angelus White before and (**D**) after rinsing and immersion of samples in simulated blood; (**E**) Biodentine before and (**F**) after rinsing and immersing the samples in simulated blood; (**G**) EndoCem MTA Zr before and (**H**) after rinsing and immersion of samples in simulated blood; (**I**) MTA HP before and (**J**) after rinsing and immersion of samples in simulated blood.

Table 2. Distribution of scores (percentages) pooled from three examiners for each material.

Experiment	Material	Score, n (%)		
		1	2	3
1	IRM [a]	45 (100)	-	-
	EndoCem Zr [a]	44 (97.78)	1 (2.22)	-
	MTA HP [a]	43 (95.56)	2 (4.44)	-
	MTA Angelus White [a]	41 (91.11)	4 (8.88)	-
	Biodentine [b]	-	13 (28.89)	32 (71.11)
2	IRM [a]	45 (100)	-	-
	EndoCem ZR [a]	45 (100)	-	-
	MTA HP [a]	44 (97.78)	1 (2.22)	-
	MTA Angelus White [a]	44 (97.78)	1 (2.22)	-
	Biodentine [b]	-	8 (17.78)	37 (82.22)

Different letters indicate significant differences between materials (a vs. b: $p < 0.0001$ for both experiments, Kruskal-Wallis test for mean score of the evaluators). No significant differences were found between materials indicated by a ($p > 0.5$ for both experiments). There were no significant differences between experiments 1 and 2 for any of the studied materials ($p > 0.3$, Mann-Whitney test).

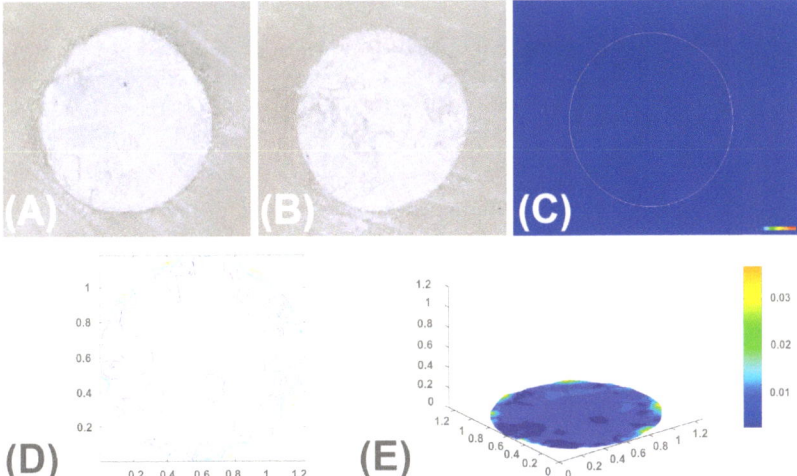

Figure 5. IRM. (**A**)—the image taken before rinsing and immersion in solution; (**B**)—the image taken after rinsing and immersion of the sample in solution—filling adheres tightly to the walls, visible loss of material indicating the washout; (**C**)—depth map of the cavity (washed out material); (**D**)—isolines showing the depth of the cavity/loss; (**E**)—3D chart of the depth of the cavity/loss.

Table 3. Interexaminer differences of score for all materials pooled.

Experiment	Comparisons	Weighted Kappa	Agreement
1	Evaluator 1 vs. 2	0.861	Very good
	Evaluator 1 vs. 3	0.890	Very good
	Evaluator 2 vs. 3	0.898	Very good
2	Evaluator 1 vs. 2	0.924	Very good
	Evaluator 1 vs. 3	0.922	Very good
	Evaluator 2 vs. 3	0.922	Very good

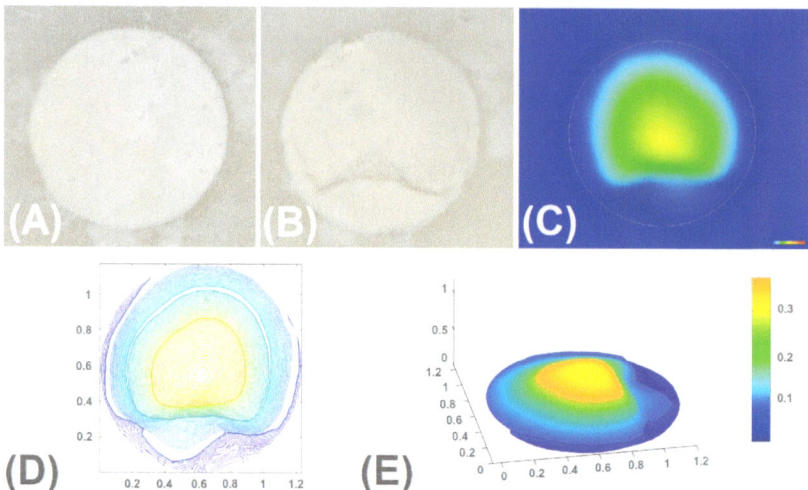

Figure 6. Biodentine. (**A**)—the image taken before rinsing and immersion in solution; (**B**)—the image taken after rinsing and immersion of the sample in the solution—filling adheres tightly to the walls, visible loss of material indicating the washout; (**C**)—depth map of the cavity (washed out material); (**D**)—isolines showing the depth of the cavity/loss; (**E**)—3D chart of the depth of the cavity/loss.

3.2. Quantitative Assessment of Washed out Material

Rinsing and immersion of the samples in a simulated blood solution resulted in a slight loss of IRM, EndoCem MTA Zr, and MTA HP. A slightly greater washout was observed with restorations made of MTA Angelus White. However, rinsing and immersion of Biodentine restorations in a simulated blood solution resulted in significant destruction. When these volumes were translated into depth changes, the mean depths of the cavity/loss within restorations made of the studied materials ranged from 0.0132 (IRM) to 0.2230 mm Biodentine). Statistical analysis showed statistically significant differences between MTA Angelus White and other materials and between Biodentine and other materials (Table 4).

Table 4. The results of the quantitative assessment of washed-out materials.

Material	Washout		
	Mean Volumetric Change ± SD (in mm^3)	Mean Volumetric Change ± SD (in %)	Mean Depth ± SD (in mm)
IRM	0.0149 [a] ± 0.0021	0.4392 [a] ± 0.0605	0.0132 [a] ± 0.0018
EndoCem Zr	0.0180 [a] ± 0.0009	0.5300 [a] ± 0.0271	0.0159 [a] ± 0.0008
MTA HP	0.0185 [a] ± 0.0029	0.5442 [a] ± 0.0885	0.0163 [a] ± 0.0026
MTA Angelus	0.0305 [b] ± 0.0089	0.9004 [b] ± 0.2627	0.0270 [b] ± 0.0079
Biodentine	0.2521 [c] ± 0.0338	7.4332 [c] ± 0.9967	0.2230 [c] ± 0.0299

Different letters indicate significant differences ($p < 0.05$, ANOVA followed by Tukey's post hoc test for log-transformed values) between materials (b vs. a: $p < 0.005$; c vs. a and c vs. b: $p = 0.00013$; no significant differences were found between three materials denoted by "a"), so the ranking of materials from best performance down is a > b > c (IRM, EndoCem Zr, MTA HP > MTA Angelus > Biodentine).

Figures 5 and 6 show the images captured by the microscope, a depth map of the cavity (washed out material), isolines showing the depth of the cavity/loss, and a 3D chart of the depth of the cavity/loss regarding the IRM (the lowest volumetric changes) and Biodentine (the highest volumetric changes).

4. Discussion

One of the desirable features of hydraulic-setting materials placed at the root end during periapical surgery is washout resistance. Therefore, newly introduced materials on the market should not show a tendency to disintegrate as a result of contact with irritants used for rinsing the resection cavity or with body fluids [5–8,47,49].

The current study aimed to compare the washout of relatively new calcium silicate cements to that of commonly used retrograde root canal materials: MTA Angels White and IRM. In order to assess the degree of washout, a model very similar to the clinical situation was used.

Visual assessment of the degree of washout of the materials rinsed immediately after placement in the retrograde cavities and 3 min after the completion of filling showed spacing defects at the material-wall interface and hollow areas on the surface of the material, essentially only in the case of the Biodentine. This confirms previous studies evaluating resistance to washout using the basket drop, a method that gives quantitative evidence of the amount of material lost when subjected to tissue fluids and irrigating solutions during the placement of root-end materials [49]. In the cited study, Biodentine washed out to a significant extent (50% of fill weight), while with radiopacified TCS cement, Bioaggregatte and IRM washed out slightly (below 10% of fill weight). The authors of the cited study explain such a high washout by the presence of a soluble polymer in the liquid, which contributed to the reduction of the water–cement ratio without affecting the workability of the resultant cement mix. In this way, it is possible to reduce the volume of water needed to mix the material and thus improve the strength properties of the cement. On the other hand, a water-soluble polymer has the effect of a surface-active agent and thus will scatter the cement particles by applying a charge to their surfaces. This scattering will lead to a fluid mixture, which results in the dislodgement of Biodentine when tested for washout [49].

In our own study, the microscopic assessment of the washout of materials showed no deterioration of marginal integrity or formation of a hollow area in the cases of EndoCem MTA, MTA HP, IRM, and Angelus MTA White. While the high resistance to washout of IRM and EndoCem MTA according to visual assessment is not surprising as this property has been demonstrated in previous studies [50], the fairly good washout resistance found in MTA Angelus White was unexpected. Partial or complete leaching of MTA was most often reported in the literature [47,50–52]. The above differences should be explained primarily by differences in the research methodology and perhaps by the fact that the currently produced MTA Angelus White differs in physical and chemical properties from the MTA Angelus White used years ago. The material currently available on the market sets within 12 min, and the one produced years ago sets within 40 min (the initial setting time). However, it is known that the tendency to wash out the material increases with the setting time of hydraulic calcium silicate cements [47]. However, to the best of our knowledge, there is no information in the literature regarding the washout resistance of the MTA HP material.

In the present study, two experiments were performed to evaluate the washout resistance of the retrograde filling materials. In the first experiment, the time between application and rinsing was 1 min (the time necessary to take a picture in the microscope); in the second experiment, a 3-min interval was used (the time necessary to take the picture and scan the surface of the restorations). Thanks to this, in the second experiment, it was possible to assess the degree of washout of the material not only visually (qualitatively) but also quantitatively. By superimposing the scan of the fillings taken before rinsing on the scan of the fillings taken after rinsing, the volume of the washed material was obtained. Quantification of the washed material confirmed the poor resistance to washout of the Biodentine as observed visually. The quantitative examination also showed a slight washout of MTA Angelus White, which was not observed visually. This observation suggests that visual evaluation allows the estimation of the washed material in cases of significant material loss. This observation suggests that visual assessment allows for the estimation of washed material in cases of significant loss of material. However, if the loss of material is insignificant, it is impossible to visually assess the loss of material under a light microscope. More precise observations can be made with a scanning microscope, which allows the depth of field yield to be assessed. Unfortunately, conventional SEM allows viewing the preparation only once (it is not possible to compare the surface of the sample before and after rinsing, as the sample is destroyed during the first test). Although environmental SEM (ESEM) or low-voltage mode of SEM operation are available where the sample can be viewed more than once (with this type of microscope/operation mode, it is not necessary to cover the viewed sample with a layer of conductive material, so it is hypothetically possible to rinse it and view it again), they cannot be used due to the time-consuming nature of the test (the sample will set before the test is performed) [53,54]. In the present study, quantitative evaluation of the washout of Endocem MTA and IRM showed minimal loss of material (within error), which is consistent with previous observations [50].

Quantitative assessment of the washed material has so far been done by mass loss. For this purpose, the material was injected into distilled water for 24 h. After this time, the sample was dried, and after comparison with the initial mass, the percentage weight loss of cement [52] was determined. Some authors placed the evaluated material in bovine serum beakers and shook them. The samples were removed from the shaker for evaluation after being shaken for 0, 5, 10, 30, and 60 min, and the percentage weight loss was assessed [43].

Another objective method of washout resistance, based on the assessment of mass loss, was described by Formosa et al. [47]. These authors adopted the test method used to estimate the resistance of freshly prepared cement to washing out in the water. This original method involves placing the studied cement into a perforated vessel, allowing it to sink freely through the water, and then raising it back up. The test cycle is repeated several times, and the mass of material washed following each cycle is estimated. The main difference between the original method and the modified one lies in the size of the samples of the studied material and the size of the device constructed for dental testing (both samples and the measuring device are correspondingly smaller) and in immersing them not in tap water but in distilled water and/or HBSS.

In the present study, a quantitative evaluation of the washed material was made by comparing the scan of the surface of the fillings registered immediately after their placement with the scan registered after rinsing and immersion in solution. A similar methodology has so far been used only by Smith et al., who used a profilometer instead of a scanner [55]. In the cited study, however, the washout was not determined, but the solubility of set calcium silicate cement in endodontic solutions (EDTA, BioPur MTAD). Fully set cements have been found to be resistant to endodontic irrigants.

The method that allows for a very accurate estimation of the lost volume of material is micro-computer tomography (micro-CT) [56]. This test allows not only the surface loss of the material but also the presence of the canal wall—retrograde filling gaps and voids within the retrograde fillings—to be assessed [57]. Micro-CT imaging is a non-invasive, highly accurate tool that has been increasingly used for the 3-dimensional assessment of microstructures. A certain limitation of the method is that only radiopaque material can be evaluated; however, the materials used for retrograde filling of the prepared canal should be radiopaque, so this is not a problem [58]. However, a significant disadvantage of this method is its time-consuming nature, which does not allow it to be used to assess washout. The scanning procedure takes so long that the material is set and, at most, its solubility but not its washout can be determined [58–60].

Many authors draw attention to the poor washout resistance of conventional MTA preparations. However, our own study did not show that MTA Angelus White was significantly washed out, which is to some extent confirmed by the good results of clinical trials [61–64]. Regarding Biodentine as a material for retrograde root canal filling, there are no randomized and prospective clinical trials in the available literature, although this material has been commercially available for 14 years. Only case reports and case series, which have limited scientific value, have been published in the literature [65–68].

The strength of the experiment was reproducibility; the authors did not have to limit themselves to a certain number of trials and could ensure the accuracy of the results. However, a weakness of the study was the relatively long time of the 3D scanning, which could not be reduced. Using a 3D scanner that can scan faster might be useful in future studies.

5. Conclusions

Under the conditions of the current study, the evaluated materials, excluding Biodentine, showed good or relatively good washout resistance. The Biodentine material was washed out both 1 and 3 min after filling, which is worrying and requires further research.

Author Contributions: Conceptualization, J.F. and M.L.; methodology, J.F. and M.L.; validation, M.L.; formal analysis, J.F., T.C. and M.L.; investigation, J.F., W.D., A.D., M.T. and E.M.; data curation, resources, J.F.; statistical analysis K.S.; writing—original draft preparation, J.F.; writing—review and editing, J.F., W.D., T.C., E.M., M.T., A.D., K.S. and M.L.; visualization, T.C. and W.D.; supervision, M.L.; project administration, J.F. and M.L. All authors have read and agreed to the published version of the manuscript.

Funding: This research received no external funding.

Institutional Review Board Statement: Not applicable.

Informed Consent Statement: Not applicable.

Data Availability Statement: Not applicable.

Conflicts of Interest: The authors declare no conflict of interest.

References

1. Kim, S.; Kratchman, S. Modern endodontic surgery concepts and practice: A review. *J. Endod.* **2006**, *32*, 601–623. [CrossRef] [PubMed]
2. Barone, C.; Dao, T.T.; Basrani, B.B.; Wang, N.; Friedman, S. Treatment outcome in endodontics: The Toronto study–phases 3, 4, and 5: Apical surgery. *J. Endod.* **2010**, *36*, 28–35. [CrossRef] [PubMed]
3. Tsesis, I.; Rosen, E.; Taschieri, S.; Telishevsky Strauss, Y.; Ceresoli, V.; Del Fabbro, M. Outcomes of surgical endodontic treatment performed by a modern technique: An updated meta-analysis of the literature. *J. Endod.* **2013**, *39*, 332–339. [CrossRef] [PubMed]
4. Von Arx, T.; Penarrocha, M.; Jensen, S. Prognostic factors in apical surgery with root-end filling: A meta-analysis. *J. Endod.* **2010**, *36*, 957–973. [CrossRef]
5. Niederman, R.; Theodosopoulou, J.N. A systematic review of in vivo retrograde obturation materials. *Int. Endod. J.* **2003**, *36*, 577–585. [CrossRef] [PubMed]
6. Johnson, R.B. Considerations in the selection of a root end filling materials. *Oral Surg. Oral Med. Oral Pathol.* **1999**, *87*, 398–404. [CrossRef]
7. Paños-Crespo, A.; Alba Sánchez-Torres, A.; Gay-Escoda, C. Retrograde filling material in periapical surgery: A systematic review. *Med. Oral Patol. Oral Cir. Bucal.* **2021**, *26*, 422–429. [CrossRef]
8. Theodosopoulou, J.N.; Niederman, R. A systematic review of in vitro retrograde obturation materials. *J. Endod.* **2005**, *31*, 341–349. [CrossRef]
9. Oleszek-Listopad, J.; Sarna-Bos, K.; Szabelska, A.; Czelej-Piszcz, E.; Borowicz, J.; Szymanska, J. The use of gold and gold alloys in prosthetic dentistry—A literature review. *Curr. Issues Pharm. Med. Sci.* **2015**, *28*, 192–195. [CrossRef]
10. Vajrabhaya, L.O.; Korsuwannawong, S.; Jantarat, J.; Korre, S. Biocompatibility of furcal perforation repair material using cell culture technique: Ketac Molar versus Pro-Root MTA. *Oral Surg. Oral Med. Oral Pathol. Oral Radiol. Endod.* **2006**, *102*, 48–50. [CrossRef]
11. Jung, S.; Mielert, J.; Kleinheinz, J.; Dammaschke, T. Human oral cells' response to different endodontic restorative materials: An in vitro study. *Head. Face. Med.* **2014**, *23*, 10–55. [CrossRef] [PubMed]
12. Torabinejad, M.; Higa, R.K.; McKendry, D.J.; Pitt Ford, T.R. Dye leakage of four root end filling materials: Effects of blood contamination. *J. Endod.* **1994**, *20*, 159–163. [CrossRef] [PubMed]
13. Katsamakis, S.; Slot, D.E.; van der Sluis, L.W.; van der Weijden, F. Histological responses of the periodontium to MTA: A systematic review. *J. Clin. Periodontol.* **2013**, *40*, 334–344. [CrossRef] [PubMed]
14. Sipert, C.R.; Hussne, R.P.; Nishiyama, C.K.; Torres, S.A. In vitro antimicrobial activity of Fill Canal, Sealapex, mineral trioxide aggregate, Portland cement and EndoRez. *Int. Endod. J.* **2005**, *38*, 539–543. [CrossRef]

15. Eldeniz, A.U.; Hadimili, H.H.; Ataoglu, H.; Ørstavik, D. Antibacterial effect of selected root-end filling materials. *J. Endod.* **2006**, *32*, 345–349. [CrossRef]
16. Baek, S.H.; Plenk, H.; Kim, S. Periapical tissue responses and cementum regeneration with amalgam, Super EBA, and MTA as root-end filling materials. *J. Endod.* **2005**, *31*, 444–449. [CrossRef]
17. Torabinejad, M.; Hong, C.U.; Lee, S.J.; Monsef, M.; Pitt Ford, T.R. Investigation of mineral trioxide aggregate for root-end filling in dogs. *J. Endod.* **1995**, *21*, 603–608. [CrossRef]
18. Torabinejad, M.; Pitt Ford, T.R.; McKendry, D.J.; Abedi, H.R.; Miller, D.A.; Kariyawasam, S.P. Histologic assessment of mineral trioxide aggregate as a root-end filling in monkeys. *J. Endod.* **1997**, *23*, 225–228. [CrossRef]
19. Duarte, M.A.H.; Marciano, M.A.; Vivan, R.R.; Tanomaru Filho, M.; Tanomaru, J.M.G.; Camilleri, J. Braz Tricalcium silicate-based cements: Properties and modifications. *Oral Res.* **2018**, *32* (Suppl. S1), 70.
20. Cervino, G.; Laino, L.; D'Amico, C.; Russo, D.; Nucci, L.; Amoroso, G.; Gorassini, F.; Tepedino, M.; Terranova, A.; Gambino, D.; et al. Mineral Trioxide Aggregate Applications in endodontics: A review. *Eur. J. Dent.* **2020**, *14*, 683–691. [CrossRef]
21. Parirokh, M.; Torabinejad, M. Mineral trioxide aggregate: A comprehensive literature review–part III: Clinical applications, drawbacks, and mechanism of action. *J. Endod.* **2010**, *36*, 400–413. [CrossRef] [PubMed]
22. Rudawska, A.; Sarna-Boś, K.; Rudawska, A.; Olewnik-Kruszkowska, E.; Frigione, M. Biological effects and toxicity of compounds based on cured epoxy resins. *Polymers* **2022**, *14*, 4915. [CrossRef] [PubMed]
23. Fagogeni, I.; Metlerska, J.; Lipski, M.; Falgowski, T.; Maciej, G.; Nowicka, A. Materials used in regenerative endodontic procedures and their impact on tooth discoloration. *J. Oral Sci.* **2019**, *61*, 379–385. [CrossRef]
24. Możynska, J.; Metlerski, M.; Lipski, M.; Nowicka, A. Tooth discoloration induced by different calcium silicate-based cements: A systematic review of in vitro studies. *J. Endod.* **2017**, *43*, 1593–1601. [CrossRef]
25. Lenherr, P.; Allgayer, N.; Weiger, R.; Filippi, A.; Attin, T.; Krastl, G. Tooth discoloration induced by endodontic materials: A laboratory study. *Int. Endod. J.* **2012**, *45*, 942–949. [CrossRef] [PubMed]
26. Kot, K.; Kucharski, Ł.; Marek, E.; Safranow, K.; Lipski, M. Alkalizing properties of six calcium-silicate endodontic biomaterials. *Materials* **2022**, *18*, 6482. [CrossRef]
27. Kaup, M.; Schäfer, E.; Dammaschke, T. An in vitro study of different material properties of Biodentine compared to ProRoot MTA. *Head Face Med.* **2015**, *2*, 11–16. [CrossRef]
28. Lipski, M.; Nowicka, A.; Kot, K.; Postek-Stefańska, L.; Wysoczańska-Jankowicz, I.; Borkowski, L.; Andersz, P.; Jarząbek, A.; Grocholewicz, K.; Sobolewska, E.; et al. Factors affecting the outcomes of direct pulp capping using Biodentine. *Clin. Oral Investig.* **2018**, *22*, 2021–2029. [CrossRef]
29. Rajasekharan, S.; Martens, L.C.; Cauwels, R.G.E.C.; Anthonappa, R.P. Biodentine™ material characteristics and clinical applications: A 3 year literature review and update. *Eur. Arch. Paediatr. Dent.* **2018**, *19*, 1–22. [CrossRef]
30. Harms, C.S.; Schäfer, E.; Dammaschke, T. Clinical evaluation of direct pulp capping using a calcium silicate cement-treatment outcomes over an average period of 2.3 years. *Clin. Oral Investig.* **2019**, *23*, 3491–3499. [CrossRef]
31. Al-Nazhan, S.; El Mansy, I.; Al-Nazhan, N.; Al-Rowais, N.; Al-Awad, G.J. Outcomes of furcal perforation management using Mineral Trioxide Aggregate and Biodentine: A systematic review. *Appl. Oral Sci.* **2022**, *30*, e20220330. [CrossRef] [PubMed]
32. Salha, M.S.; Yada, R.Y.; Farrar, D.H.; Chass, G.A.; Tian, K.V.; Bodo, E. Aluminium catalysed oligomerisation in cement-forming silicate systems. *Phys. Chem. Chem. Phys.* **2023**, *25*, 455–461. [CrossRef] [PubMed]
33. Silva, E.J.N.L.; Carvalho, N.K.; Guberman, M.R.D.C.L.; Prado, M.; Senna, P.M.; Souza, E.M.; De-Deus, G. Push-out bond strength of fast-setting Mineral Trioxide Aggregate and pozzolan-based cements: ENDOCEM MTA and ENDOCEM Zr. *J. Endod.* **2017**, *43*, 801–804. [CrossRef]
34. Sharma, A.; Thomas, M.S.; Shetty, N.; Srikant, N.J. Evaluation of indirect pulp capping using pozzolan-based cement (ENDOCEM-Zr®) and mineral trioxide aggregate—A randomized controlled trial. *Conserv. Dent.* **2020**, *23*, 152–157.
35. Kim, M.; Yang, W.; Kim, H.; Ko, H. Comparison of the biological properties of ProRoot MTA, OrthoMTA, and Endocem MTA cements. *J. Endod.* **2014**, *40*, 1649–1653. [CrossRef] [PubMed]
36. Choi, Y.; Park, S.J.; Lee, S.H.; Hwang, Y.C.; Yu, M.K.; Min, K.S. Biological effects and washout resistance of a newly developed fast-setting pozzolan cement. *J. Endod.* **2013**, *39*, 467–472. [CrossRef]
37. Chung, C.J.; Kim, E.; Song, M.; Park, J.W.; Shin, S.J. Effects of two fast-setting calcium-silicate cements on cell viability and angiogenic factor release in human pulp-derived cells. *Odontology* **2016**, *104*, 143–151. [CrossRef] [PubMed]
38. Lee, M.; Kang, C.M.; Song, J.S.; Shin, Y.; Kim, S.; Kim, S.O.; Choi, H.J. Biological efficacy of two mineral trioxide aggregate (MTA)-based materials in a canine model of pulpotomy. *Dent. Mater. J.* **2017**, *31*, 41–47. [CrossRef]
39. Palczewska-Komsa, M.; Kaczor-Wiankowska, K.; Nowicka, A. New bioactive calcium silicate cement mineral Trioxide Aggregate Repair High Plasticity (MTA HP)—A systematic review. *Materials* **2021**, *14*, 4573. [CrossRef]
40. Barczak, K.; Palczewska-Komsa, M.; Lipski, M.; Chlubek, D.; Buczkowska-Radlinska, J.; Baranowska-Bosiacka, I. The influence of new silicate cement Mineral Trioxide Aggregate (MTA Repair HP) on metalloproteinase MMP-2 and MMP-9 expression in cultured THP-1 macrophages. *Int. J. Mol. Sci.* **2020**, *22*, 295. [CrossRef]
41. Galarça, A.D.; Rosa, W.L.D.O.D.; Da Silva, T.M.; Lima, G.D.S.; Carreño, N.L.V.; Pereira, T.M.; Guedes, O.A.; Borges, A.H.; da Silva, A.F.; Piva, E. Physical and biological properties of a high-plasticity tricalcium silicate cement. *BioMed Res. Int.* **2018**, *2018*, 8063262. [CrossRef]

42. Jimenez-Sanchez, M.; Segura-Egea, J.; Diaz-Cuenca, A. Physicochemical parameters-hydration performance relationship of the new endodontic cement MTA Repair HP. *J. Clin. Exp. Dent.* **2019**, *11*, 739–744. [CrossRef]
43. Wang, X.; Chen, L.; Xiang, H.; Ye, J. Influence of anti-washout agents on the rheological properties and injectability of a calcium phosphate cement. *J. Biomed. Mater. Res.* **2007**, *81*, 410–418. [CrossRef]
44. Megahed, A.E.; Adam, I.; Farghal, O.; Sayed, M.O. Influence of mixture composition on washout resistance, fresh properties and relative strength of self compacting underwater concrete. *J. Eng. Sci.* **2016**, *44*, 244–258. [CrossRef]
45. Jang, G.Y.; Park, S.J.; Heo, S.M.; Yu, M.K.; Lee, K.W.; Min, K.S. Washout resistance of fast-setting pozzolan cement under various root canal irrigants. *Restor. Dent. Endod.* **2013**, *38*, 248–252. [CrossRef]
46. Sonebi, M.; Bartos, P.J.M.; Khayat, K.H. Assessment of washout resistance of underwater concrete: A comparison between CRD C61 and new MC-1 tests. *Mater. Struct.* **1999**, *32*, 273. [CrossRef]
47. Formosa, L.M.; Mallia, B.; Camilleri, J. A quantitative method for determining the anti-washout characteristics of cement-based dental materials including mineral trioxide aggregate. *Int. Endod. J.* **2013**, *46*, 179–186. [CrossRef]
48. Tran, D.; He, J.; Glickman, G.N.; Woodmansey, K.F. Comparative analysis of calcium silicate-based root filling materials using an open apex model. *J. Endod.* **2016**, *42*, 654–658. [CrossRef]
49. Grech, L.B.; Mallia, J.; Camilleri, J. Investigation of the physical properties of tricalcium silicate cement-based root-end filling materials. *Dent Mater.* **2013**, *29*, 20–28. [CrossRef]
50. Porter, M.L.; Bertó, A.; Primus, C.M.; Watanabe, I. Physical and chemical properties of new-generation endodontic materials. *J. Endod.* **2010**, *36*, 524–528. [CrossRef]
51. Bortoluzzi, E.A.; Broon, N.J.; Bramante, C.M.; Garcia, R.B.; de Moraes, I.G.; Bernardineli, N. Sealing ability of MTA and radiopaque Portland cement with or without calcium chloride for root-end filling. *J. Endod.* **2006**, *32*, 897–900. [CrossRef] [PubMed]
52. Manero, J.M.; Gil, F.J.; Padrós, E.; Planell, J.A. Applications of environmental scanning electron microscopy (ESEM) in biomaterials field. *Microsc. Res. Tech.* **2003**, *61*, 469–480. [CrossRef]
53. Athene, M.D. The use of environmental scanning electron microscopy for imaging wet and insulationg materials. *Nat. Mater.* **2003**, *2*, 511–516.
54. Kai, D.; Li, D.; Zhu, X.; Zhang, L.; Fan, H.; Zhang, X. Addition of sodium hyaluronate and the effect on performance of the injectable calcium phosphate cement. *J. Mater. Sci.* **2009**, *20*, 1595–1602. [CrossRef]
55. Smith, J.B.; Loushine, R.J.; Weller, R.N.; Rueggeberg, F.A.; Whitford, G.M.; Pashley, D.H.; Tay, F.R. Metrologic evaluation of the surface of White MTA after the use of two endodontic irrigants. *J. Endod.* **2007**, *33*, 463–467. [CrossRef]
56. Kim, S.Y.; Kim, H.C.; Shin, S.J.; Kim, E. Comparison of gap volume after retrofilling using 4 different filling materials: Evaluation by micro–computed tomography. *J. Endod.* **2018**, *44*, 635–638. [CrossRef] [PubMed]
57. Jung, J.; Kim, S.; Kim, E.; Shin, S.J. Volume of voids in retrograde filling: Comparison between calcium silicate cement alone and combined with a calcium silicate–based sealer. *J. Endod.* **2020**, *46*, 97–102. [CrossRef]
58. Zordan-Bronzel, C.L.; Tanomaru-Filho, M.; Torres, F.F.E.; Chávez-Andrade, G.M.; Rodrigues, E.M.; Guerreiro-Tanomaru, J.M. Physicochemical properties, cytocompatibility and antibiofilm activity of a new calcium silicate sealer. *Braz. Dent. J.* **2021**, *32*, 8–18. [CrossRef]
59. Torres, F.F.E.; Zordan-Bronzel, C.L.; Guerreiro-Tanomaru, J.M.; Chávez-Andrade, G.M.; Pinto, J.C.; Tanomaru-Filho, M. Effect of immersion in distilled water or phosphate-buffered saline on the solubility, volumetric change and presence of voids within new calcium silicate-based root canal sealers. *Int. Endod. J.* **2020**, *53*, 385–391. [CrossRef]
60. Zordan-Bronzel, C.L.; Esteves Torres, F.F.; Tanomaru-Filho, M.; Chávez-Andrade, G.M.; Bosso-Martelo, R.; Guerreiro-Tanomaru, J.M. Evaluation of physicochemical properties of a new calcium silicate-based sealer, Bio-C Sealer. *J. Endod.* **2019**, *45*, 1248–1252. [CrossRef]
61. Chong, B.S.; Pitt Ford, T.R.; Hudson, M.B. A prospective clinical study of mineral trioxide aggregate and IRM when used as root-end filling materials in endodontic surgery. *Int. Endod. J.* **2003**, *36*, 520–526. [CrossRef] [PubMed]
62. Lindeboom, J.A.; Frenken, J.W.; Kroon, F.H.; van den Akker, H.P. A comparative prospective randomized clinical study of MTA and IRM as root-end filling materials in single-rooted teeth in endodontic surgery. *Oral Surg. Oral Med. Oral Pathol. Oral Radiol. Endod.* **2005**, *100*, 495–500. [CrossRef] [PubMed]
63. Christiansen, R.; Kirkevang, L.L.; Hørsted-Bindslev, P.; Wenzel, A. Randomized clinical trial of root-end resection followed by root-end filling with mineral trioxide aggregate or smoothing of the orthograde gutta-percha root filling 1-year follow-up. *Int. Endod. J.* **2009**, *42*, 105–114. [CrossRef] [PubMed]
64. Kruse, C.; Spin-Neto, R.; Christiansen, R.; Wenzel, A.; Lise-Lotte, K. Periapical bone healing after apicectomy with and without retrograde root filling with mineral trioxide aggregate: A 6-year follow-up of a randomized controlled trial. *J. Endod.* **2016**, *42*, 533–537. [CrossRef]
65. Caron, G.; Azérad, J.; Faure, M.O.; Machtou, P.; Boucher, Y. Use of a new retrograde filling material (Biodentine) for endodontic surgery: Two case reports. *Int. J. Oral Sci.* **2014**, *6*, 250–253. [CrossRef]
66. Pawar, A.M.; Kokate, S.R.; Shah, R.A. Management of a large periapical lesion using Biodentine as retrograde restoration with eighteen months evident follow up. *J. Conserv. Dent.* **2013**, *16*, 573–575. [CrossRef]

67. Wälivaara, D.A. Periapical surgery with Biodentine™ as a retrograde root-end seal: A clinical case series study. *Med. Mat. Sci.* **2015**, *14*, 120–124.
68. Bhagwat, S.; Hegde, S.; Mandke, L. An investigation into the effectiveness of periapical surgery with Biodentine™ used as a root-end filling alone or in combination with demineralized freeze-dried bone allograft and plasma rich fibrin: A 6 months follow-up of 17 cases. *Endodontology* **2016**, *28*, 11–17. [CrossRef]

Disclaimer/Publisher's Note: The statements, opinions and data contained in all publications are solely those of the individual author(s) and contributor(s) and not of MDPI and/or the editor(s). MDPI and/or the editor(s) disclaim responsibility for any injury to people or property resulting from any ideas, methods, instructions or products referred to in the content.

Article

Assessment of pH Value and Release of Calcium Ions in Calcium Silicate Cements: An In Vitro Comparative Study

Rubén Herrera-Trinidad [1,*], Pedro Molinero-Mourelle [1,2], Manrique Fonseca [2], Adrian Roman Weber [2], Vicente Vera [3], María Luz Mena [4] and Vicente Vera-González [1]

1. Department of Conservative Dentistry and Orofacial Prosthodontics, Faculty of Dentistry, Complutense University of Madrid, 28040 Madrid, Spain
2. Department of Reconstructive Dentistry and Gerodontology, School of Dental Medicine, University of Bern, 3007 Bern, Switzerland
3. Department of Prosthodontics and Operative Dentistry, School of Dental Medicine, Tufts University, Boston, MA 02111, USA
4. Department of Analytics Chemistry, Faculty of Chemical Sciences, Complutense University of Madrid, 28040 Madrid, Spain
* Correspondence: ruben.herrera@ucm.es

Citation: Herrera-Trinidad, R.; Molinero-Mourelle, P.; Fonseca, M.; Weber, A.R.; Vera, V.; Mena, M.L.; Vera-González, V. Assessment of pH Value and Release of Calcium Ions in Calcium Silicate Cements: An In Vitro Comparative Study. *Materials* 2023, *16*, 6213. https://doi.org/10.3390/ma16186213

Academic Editors: Maria Francesca Sfondrini, Luigi Generali, Vittorio Checchi and Eugenio Pedullà

Received: 17 July 2023
Revised: 7 September 2023
Accepted: 11 September 2023
Published: 14 September 2023

Copyright: © 2023 by the authors. Licensee MDPI, Basel, Switzerland. This article is an open access article distributed under the terms and conditions of the Creative Commons Attribution (CC BY) license (https://creativecommons.org/licenses/by/4.0/).

Abstract: The goal of this study was to evaluate the pH and the release of calcium from four calcium-silicate-based cements. Methods: Four materials were tested (ProClinic MTA; Angelus MTA; ProRoot MTA; Biodentine). The palatal canal root of acrylic upper molars was filled with each cement. Afterwards, they were set in phosphate-buffered saline. Measurements were taken by atomic adsorption spectroscopy (AAS) at 3, 24, 72, 168, 336, 672, and 1008 h. The pH was measured at the same timepoints. Kruskal–Wallis tests were carried out in each period, as the Kolmogorov–Smirnov and Shapiro–Wilk tests showed no parametric results. Results: Significant differences ($p < 0.05$) in calcium release were found at the 3-, 24-, and 72-hour evaluations. All of the analyzed groups presented a release of calcium ions up to 168 h, and the general tendency was to increase up to 672 h, with a maximum release of 25.45 mg/g in the ProRoot group. We could only observe significant differences ($p < 0.05$) in pH value over 168 h between the Biodentine (7.93) and Angelus MTA (7.31) groups. Conclusions: There were significant differences ($p < 0.05$) in calcium release. Nevertheless, no significant differences ($p > 0.05$) in the pH values were found at the studied timepoints, except for the values at 168 h.

Keywords: biomaterial; ion release; pH; mineral trioxide aggregate; MTA; calcium-silicate-based cements

1. Introduction

Calcium silicate cements (CSCs) mostly comprise dicalcium silicate (Ca_2SiO_4) and tricalcium silicate (Ca_3SiO_5). These particles primarily release calcium hydroxide and calcium silicate hydrate gel via a hydration reaction [1]. The first historical record of the use of a calcium silicate cement in dentistry was in 1878, when the German dentist Dr. Witte used Portland cement to fill dental root canals [2]. However, it was not until 1993 that Torabinejad described mineral trioxide aggregate (MTA) [3]. Later, in 1995, it was patented after the addition of bismuth powder [4].

MTA is indicated for apical fillings, direct pulp capping, canal perforation repair, and apexification [5,6]. However, MTA has some limitations to its use, such as working time, difficulty to manipulate, discoloration, and high economic cost [7]. In order to overcome these limitations, Biodentine™ has been developed; this material is also based on calcium silicates [8]. Both mineral trioxide aggregate (MTA) and Biodentine™ are materials that are combinations of dicalcium and tricalcium silicates, together with metal ions [9,10].

Calcium and hydroxyl ions are the main chemical components released by calcium silicate cements in water [3]. The release of these ions is responsible for these materials having certain properties, among which are the differentiation of pulp cells, cementoblasts, osteoblasts, periodontal fibroblasts, mesenchymal stem cells, and hard-tissue mineralization [11–15]. Calcium silicate cements allow for the angiogenic differentiation of pulp cells. In fact, the ionic products of calcium silicates induce osteogenesis and angiogenesis [16].

Alkaline pH values accelerate apatite nucleation, because OH ions become soluble and can be included in apatite—an essential component of the tooth matrix. In addition, hydroxyl ions stimulate the release of alkaline phosphatase and bone morphogenetic protein 2 (BMP2), which are involved in mineralization processes [17]. These processes of mineralization of the hard tissue and formation of apatite are responsible for the sealing of these materials and, therefore, for their indications.

There are currently few studies comparing the pH and the release of calcium ions in the medium between different types of calcium silicate and bioceramic cements for periods longer than seven days. The release after this period is an important question, since the bone matrix crystallization occurs on the surface of the bioceramic material seven days after injury [18].

According to the study by Natale et al., all of the materials decrease their calcium release at 28 days [19]. Due to this, we carried out the last measurement of the variables at 42 days. Previous studies have shown similar results, like those obtained by Cavenago et al. [20], Gandolfi et al. [11], Kim et al. [21], Bernardi et al. [22], and Irawan et al. [18], who expressed the results with a heterogeneous unit of measurement, making comparison with other studies difficult. Considering this heterogeneity limitation, further studies are necessary to allow a direct comparison.

Therefore, the aim of the present study was to evaluate the release of calcium from four calcium-silicate-based cements, along with the pH reached, to determine which material presents the highest calcium release and pH changes. The null hypothesis was that there would be no differences in the release of calcium and the pH reached among the calcium-silicate-based cements.

2. Materials and Methods

2.1. Study Design

An in vitro comparative study was performed at the Department of Conservative Dentistry and Orofacial Prosthodontics based on the previously reported protocols of Cavenago et al. [20], Gandolfi et al. [17], and Irawan et al. [18], evaluating 40 standardized acrylic teeth ("Moulding Molar/Moulding Root" by PolyJet printing. Prototype: 0076209; AIJU, Ibi, Alicante, Spain) with root canals, and an apical stop was created with an 80 file (K-Flexofile ISO 80; DentsplyTulsa Dental Specialties, Tulsa, OK, USA; REF: A012D03108004). Since no human samples were used, ethical approval was not required from the Ethics Committee Research of Complutense University Hospitals for this in vitro study.

2.2. Study Setup and Specimen Fabrication

The teeth were randomly distributed into 4 groups ($n = 10$), in which the following materials were assessed: ProClinic MTA (ProClinic, Maruchi, Wonju-si, Republic of Korea), Angelus MTA (Angelus, Londrina, PR, Brasil), ProRoot MTA (DentsplyTulsa Dental Specialties, Tulsa, OK), and Biodentine (Septodont, Saint Maurde Fossés, France).

Acrylic teeth were weighed on a previously calibrated high-precision laboratory balance (Sartorius Secura® Analytical Balance 220 g 0.1 mg. Item no.: SECURA224-1CEU). Afterwards, the instrumented canal was filled in its last three millimeters with the cement corresponding to the group, using an MTA transporter instrument (CHL MEDICAL SOLUTIONS, Srl., Milan, Italy; REF: 59814), and they were weighed again to control the amount of material introduced into the tooth. Subsequently, the teeth were placed individually in tubes with 10 mL of a buffered solution (phosphate-buffered saline, batch number 120438, Condalab, Madrid, Spain) that resembled the physiological medium [17]. In addition, four tubes were prepared as negative controls: two tubes containing one undrilled acrylic tooth, and two tubes without acrylic teeth. All of the included tubes were stored in a 37 °C distilled water bath throughout the experiment.

The calcium release determination was performed by atomic absorption spectroscopy (AAS) using an acetylene–air flame and equipped with a specific calcium-ion cathode lamp (Atomic Absorption Spectrophotometer Perkins Model 3100. PerkinElmer Inc., Waltham, MA, USA). An external calibration was performed in 25 mL flasks, and an additional control flask without calcium was prepared for the instrumental calibration. The absorbance was measured at 422.7 nm, and the calcium concentration was determined by interpolating its value in the previous calibration.

The pH determination was performed with a pH meter (Crison Basic 20+, CRISON Instruments, S.A., Alella, Barcelona, Spain) previously calibrated using buffer solutions of pH 4.7 and 9. Calcium ion release and pH measurements were performed at the following six timepoints on samples and controls: 24 h, 72 h, 168 h, 336 h, 672 h, and 1008 h.

SEM analysis using a JEOL JSM 6335F microscope and X-ray diffraction were carried out to determine the composition of each cement and to conduct a descriptive evaluation of the cements at 1008 h, taking a sample from each group.

2.3. Statistical Analysis

Data analysis was performed by using SPSS 28.0.1.0 statistical software (IBM Corp, Armonk, NY, USA). Kruskal–Wallis tests were carried out in each period, as the Kolmogorov–Smirnov and Shapiro–Wilk tests showed no parametric results. The level of significance was set to 0.05.

3. Results

The results showed an increase in calcium release and pH changes, especially over the first week (168 h). The pH values increased until reaching 7.93 in the Biodentine group at 168 h, which was the highest value in our study. In the rest of the groups the pH reached 7.48–7.51 at 24 h and then decreased.

All of the analyzed groups presented a release of calcium ions up to 7 days (168 h), and the general tendency was to increase up to 28 days. By that time, the ProClinic and Angelus groups decreased their release of calcium ions, while the ProRoot and Biodentine groups continued to increase their ion release. The pH value increased until reaching 7.93 in the Biodentine group. Significant differences ($p < 0.05$) were noticed over 168 h between the BD (7.93) and AN (7.31) groups. Medians and quantiles were also calculated due to nonparametric data (Tables 1 and 2; Figures 1–8).

Table 1. Means and standard deviations of released calcium ion concentrations (mg/g) and pH values over time; (a,b,x) show significant differences between group pairs.

	ProClinic Group (PC) $n = 10$		Angelus Group (AN) $n = 10$		ProRoot Group (PR) $n = 10$		Biodentine Group (BD) $n = 10$	
	$[Ca^{2+}]$ mg/g	pH	$[Ca^{2+}]$ mg/g	pH	$[Ca^{2+}]$ mg/g	pH	$[Ca^{2+}]$ mg/g	pH
3 h	4.44 ± 1.08 (a)	7.43 ± 0.01	2.95 ± 0.82 (a)	7.41 ± 0.01	3.36 ± 1.19	7.41 ± 0.02	3.36 ± 1.22	7.48 ± 0.12
24 h	5.37 ± 1.35 (b)	7.51 ± 0.01	4.36 ± 0.81 (a)	7.48 ± 0.01	3.66 ± 1.64	7.50 ± 0.03	2.61 ± 0.37 (a,b)	7.75 ± 0.19
72 h	4.71 ± 1.22 (a)	7.39 ± 0.04	4.57 ± 1.91 (a)	7.38 ± 0.02	6.64 ± 2.89 (a)	7.40 ± 0.04	4.28 ± 0.99 (a)	7.73 ± 0.04
168 h	10.28 ± 3.17	7.35 ± 0.07	9.15 ± 3.80	7.31 ± 0.02 (x)	13.48 ± 5.74	7.39 ± 0.07	14.80 ± 9.03	7.93 ± 0.02 (x)
336 h	13.11 ± 5.10	7.12 ± 0.02	11.03 ± 2.49	7.13 ± 0.09	13.11 ± 6.13	7.22 ± 0.03	12.86 ± 6.32	7.76 ± 0.09
672 h	13.73 ± 3.97	7.05 ± 0.01	12.52 ± 3.92 (a)	7.10 ± 0.14	25.45 ± 18.63 (a)	7.28 ± 0.16	17.58 ± 6.99	7.67 ± 0.03
1008 h	12.31 ± 3.64	7.05 ± 0.01	9.35 ± 2.23 (a,b)	7.10 ± 0.14	23.83 ± 12.45 (a)	7.28 ± 0.16	20.07 ± 5.41 (b)	7.67 ± 0.03

Note: a, b, x; denote a subset of groups whose column proportions do not differ significantly from one another at the 0.05 level, with Bonferroni's correction.

Table 2. Medians and quantiles of released ion calcium concentrations (mg/g) and pH values over time; (a,b,x) show significant differences between group pairs.

	ProClinic Group (PC) $n = 10$		Angelus Group (AN) $n = 10$		ProRoot Group (PR) $n = 10$		Biodentine Group (BD) $n = 10$	
	[Ca^{2+}] mg/g	pH	[Ca^{2+}] mg/g	pH	[Ca^{2+}] mg/g	pH	[Ca^{2+}] mg/g	pH
3 h	4.20 (1.44) (a)	7.43 (0.02)	2.79 (1.66) (a)	7.41 (0.01)	3.29 (1.92)	7.41 (0.03)	3.39 (1.69)	7.48 (0.17)
24 h	4.96 (2.50) (b)	7.51 (0.01)	4.44 (1.23) (a)	7.48 (0.01)	3.08 (1.44)	7.5 (0.05)	2.60 (0.67) (a,b)	7.75 (0.27)
72 h	4.71 (2.66) (a)	7.39 (0.06)	4.31 (1.44) (a)	7.38 (0.03)	6.16 (5.31) (a)	7.4 (0.06)	4.34 (1.67) (a)	7.73 (0.06)
168 h	9.54 (3.62)	7.35 (0.10)	8.55 (5.99)	7.31 (0.03)	14.72 (7.10)	7.39 (0.10)	14.27 (16.10)	7.93 (0.03) (x)
336 h	12.40 (5.68)	7.12 (0.04)	10.71 (4.88)	7.13 (0.14)	12.12 (10.72)	7.22 (0.05)	11.18 (5.29)	7.76 (0.13)
672 h	13.15 (7.05)	7.05 (0.01)	11.73 (4.84) (a)	7.1 (0.21)	19.85 (14.91) (a)	7.28 (0.23)	15.91 (10.73)	7.67 (0.05)
1008 h	11.37 (3.64)	7.05 (0.01)	9.07 (4.43) (a,b)	7.1 (0.21)	19.00 (21.85) (a)	7.28 (0.23)	19.09 (10.10) (b)	7.67 (0.05)

Note: a, b, x; denote a subset of groups whose column proportions do not differ significantly from one another at the 0.05 level, with Bonferroni's correction.

At 3 h, there were significant differences ($p < 0.05$) in calcium release, where the ProClinic group reached 4.40 mg/g, compared to 2.95 mg/g in the group Angelus. Over 24 h, both the ProClinic group (with 5.37 mg/g) and the Angelus group (with 4.36 mg/g) had significant differences ($p < 0.05$) from the Biodentine group (with 2.61 mg/g). At 72 h, the ProClinic group released 4.71 mg/g, the Angelus group released 4.57 mg/g, the ProRoot group released 6.64 mg/g, and the Biodentine group released 4.28 mg/g. There were significant differences ($p < 0.05$) between the groups PC–PR, PC–BD, BD–AN, and AN–PR. All analyzed groups presented a release of calcium ions up to 168 h, and the general tendency was to increase up to 672 h, with a maximum release of 25.45 mg/g in the PR group. We observed a general increase in calcium release (in mg/L) in all of the groups, where the BD group kept increasing with a higher concentration than the other groups (Table 3; Figure 9). However, the data were not statistically analyzed.

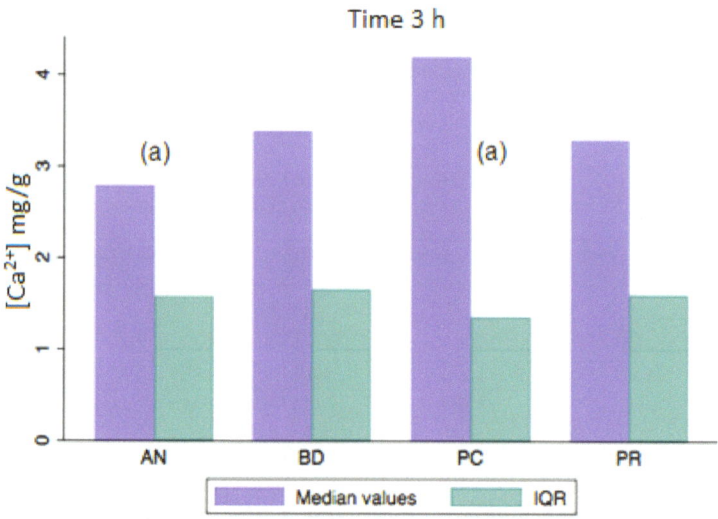

Figure 1. Released calcium ion concentration medians at 3 h, in mg/g; (a) shows significant differences between group pairs.

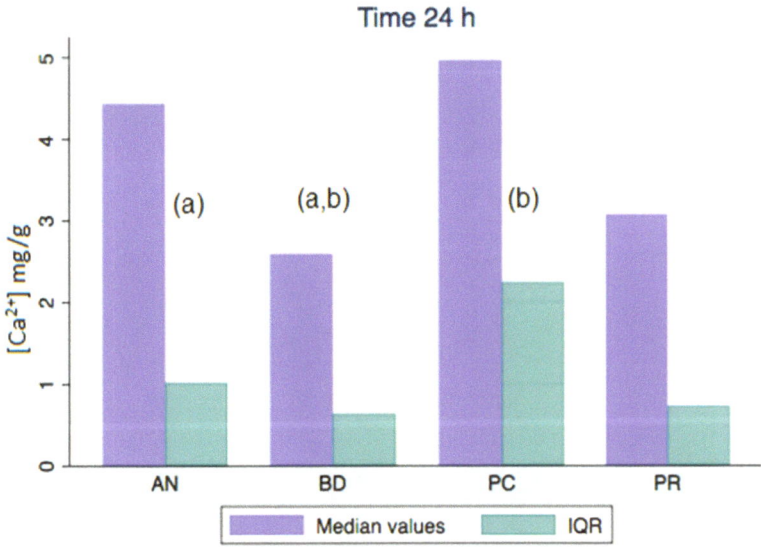

Figure 2. Released calcium ion concentration medians at 24 h, in mg/g; (a,b) show significant differences between group pairs.

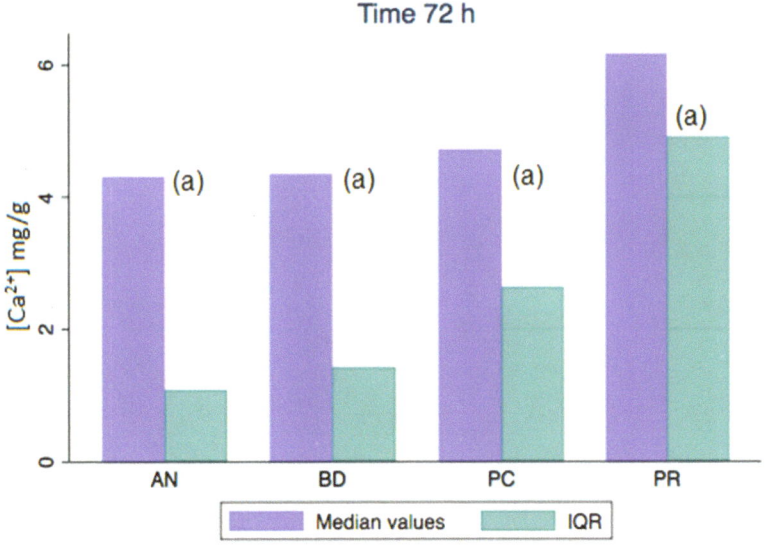

Figure 3. Released calcium ion concentration medians at 72 h, in mg/g; (a) shows significant differences between group pairs.

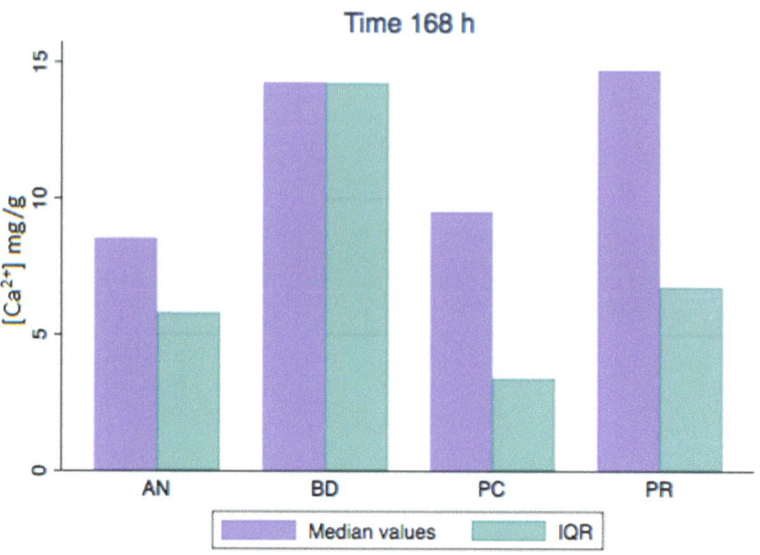

Figure 4. Released calcium ion concentration medians at 168 h, in mg/g.

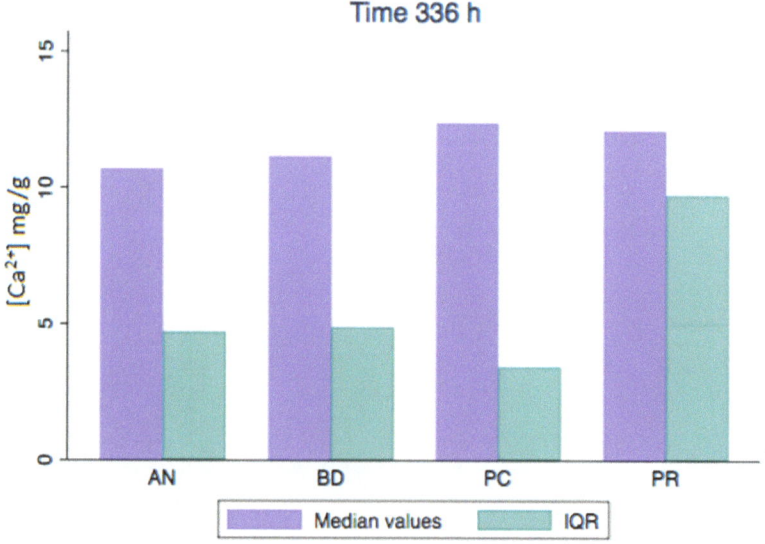

Figure 5. Released calcium ion concentration medians at 336 h, in mg/g.

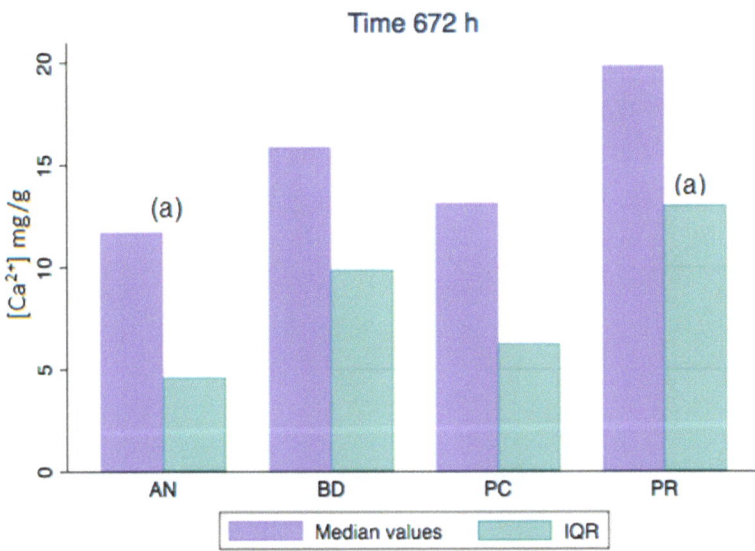

Figure 6. Released calcium ion concentration medians at 672 h, in mg/g; (a) show significant differences between group pairs.

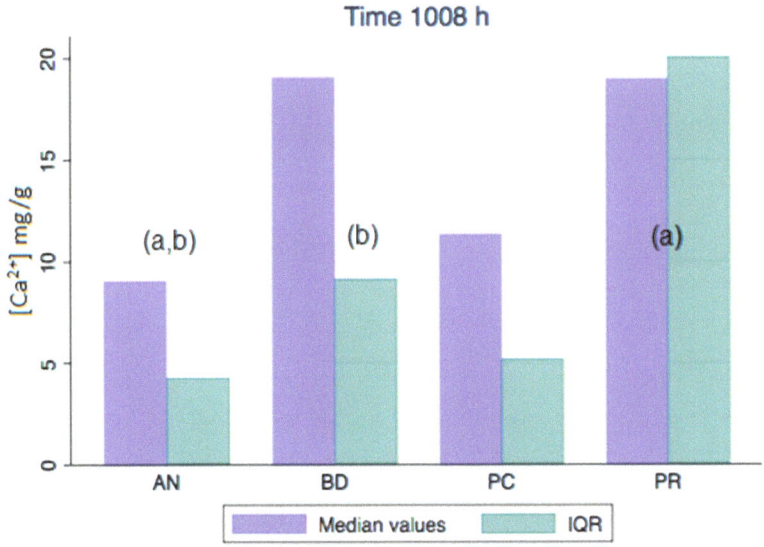

Figure 7. Released calcium ion concentration medians at 1008 h, in mg/g; (a,b) show significant differences between group pairs.

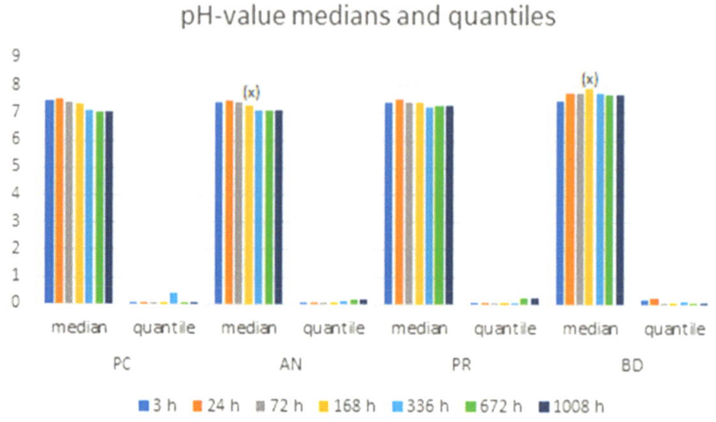

Figure 8. Medians and quantiles of pH over time; (x) shows significant differences between group pairs.

Table 3. Mean released calcium ion concentrations, measured in mg/L.

	[Ca^{2+}], mg/L			
	ProClinic Group (PC)	Angelus Group (AN)	ProRoot Group (PR)	Biodentine Group (BD)
3 h	4.76	3.61	4.98	7.20
24 h	6.34	5.82	5.84	9.15
72 h	6.22	6.52	12.24	17.06
168 h	15.01	15.23	27.10	56.95
336 h	20.16	19.96	26.39	50.15
672 h	23.00	24.95	52.87	71.00
1008 h	21.94	19.57	51.13	85.95

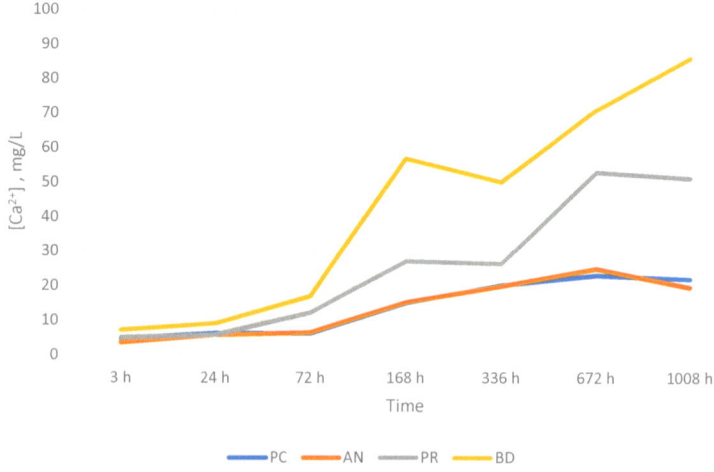

Figure 9. Mean released calcium ion concentrations, measured in mg/L.

SEM analysis using a JEOL JSM 6335F microscope and X-ray diffraction were carried out in order to determine the present phases and chemical composition of each cement. In this sense, a descriptive evaluation of the cements was conducted at 1008 h, taking a sample from each group. It was determined that all of the cements had similar chemical compositions, and they were identified as calcium aluminosilicates ($CaAl_2Si_6O_{16}$) (Figures 10–18).

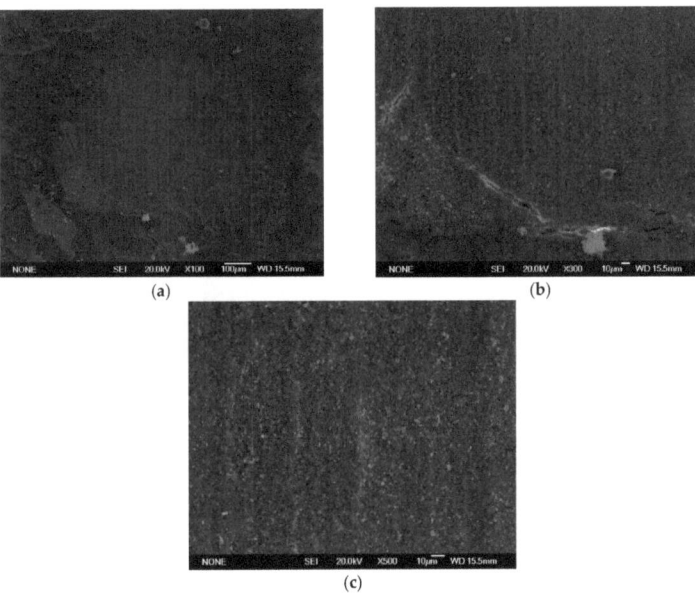

Figure 10. (a–c) Group PC sample viewed by SEM at ×100, ×300, and ×500, respectively, after 1008 h.

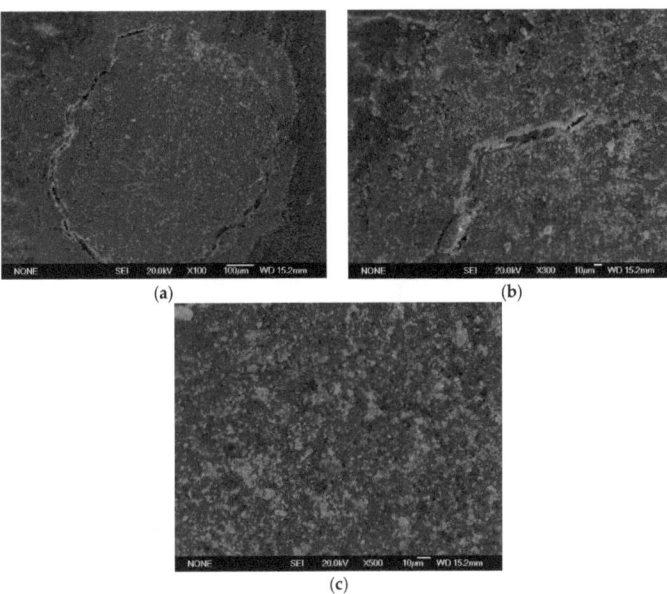

Figure 11. (a–c) Group AN sample viewed by SEM at ×100, ×300, and ×500, respectively, after 1008 h.

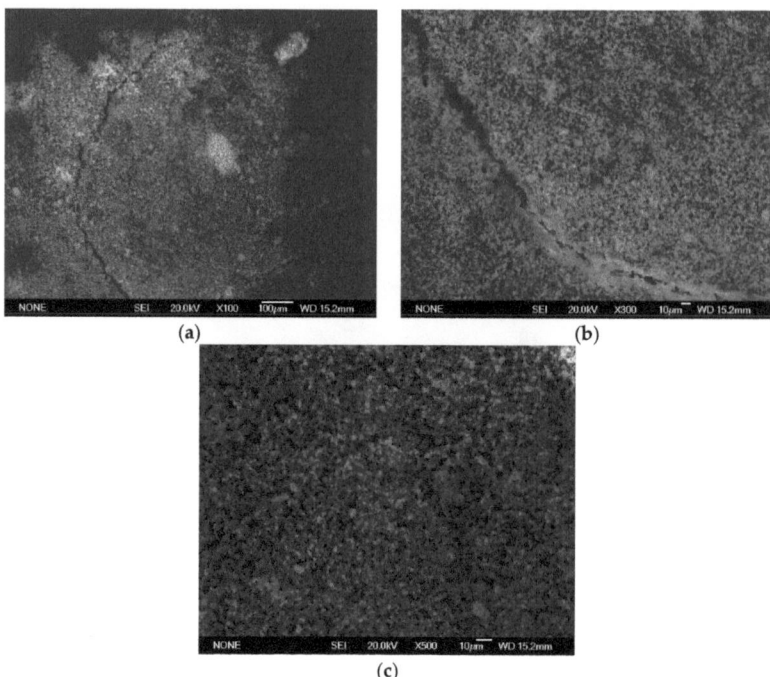

Figure 12. (**a**–**c**) Group PR sample viewed by SEM at ×100, ×300, and ×500, respectively, after 1008 h.

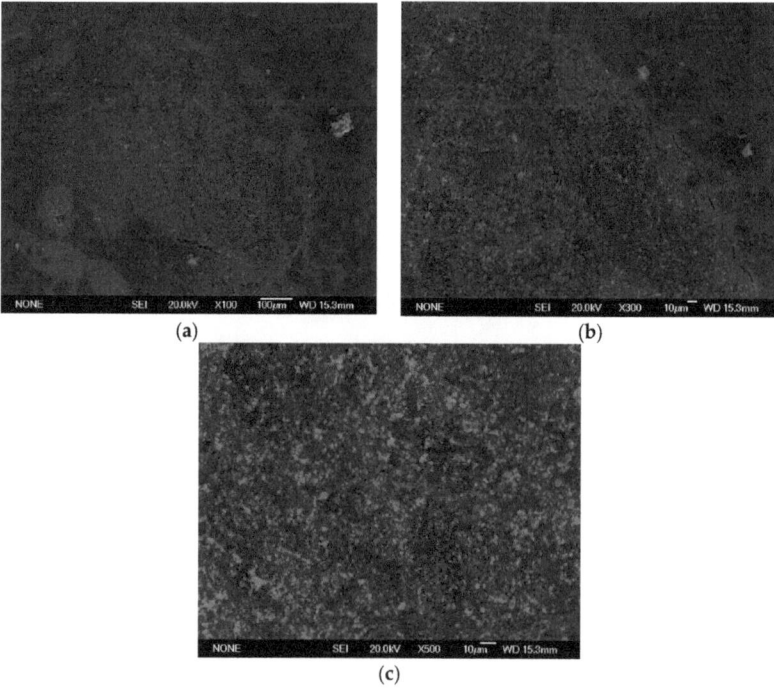

Figure 13. (**a**–**c**) Group BD sample viewed by SEM at ×100, ×300, and ×500, respectively, after 1008 h.

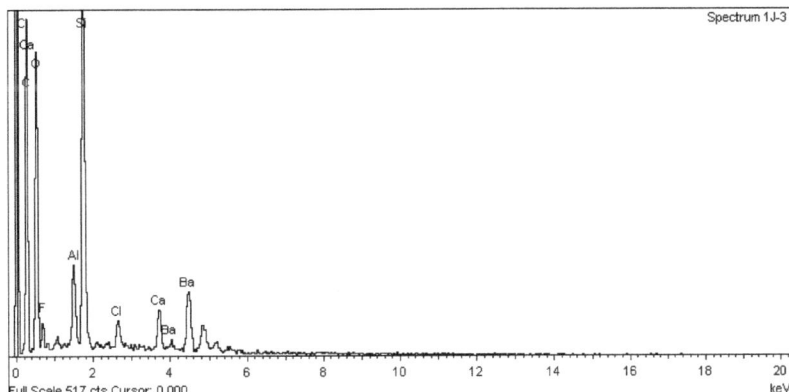

Figure 14. EDX spectrum from a PC sample.

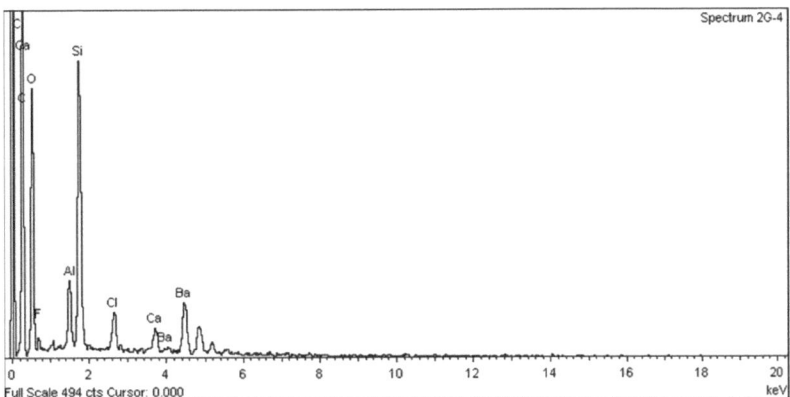

Figure 15. EDX spectrum from an AN sample.

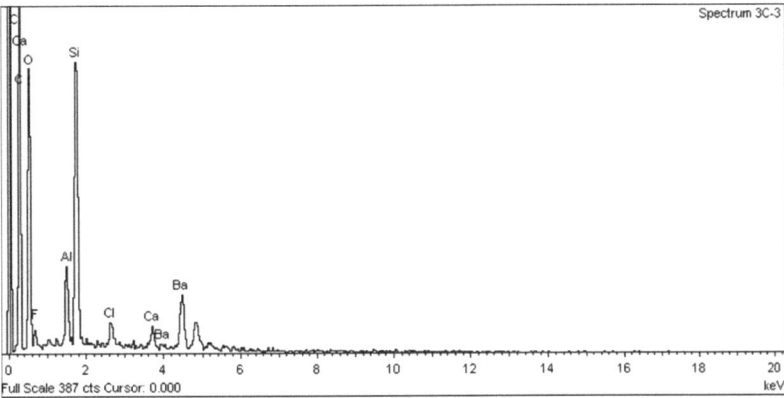

Figure 16. EDX spectrum from a PR sample.

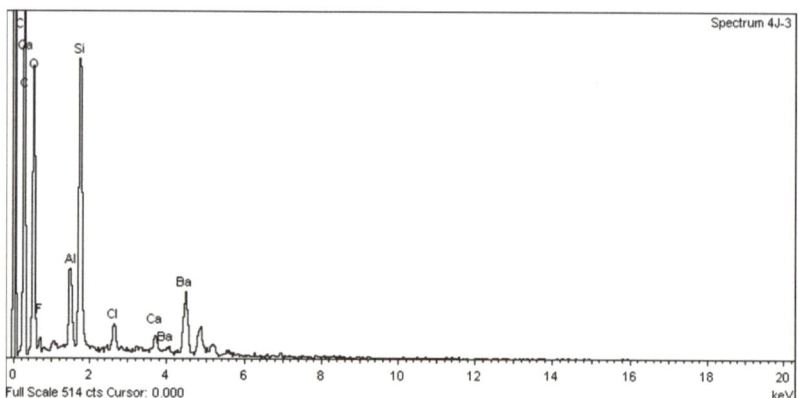

Figure 17. EDX spectrum from a BD sample.

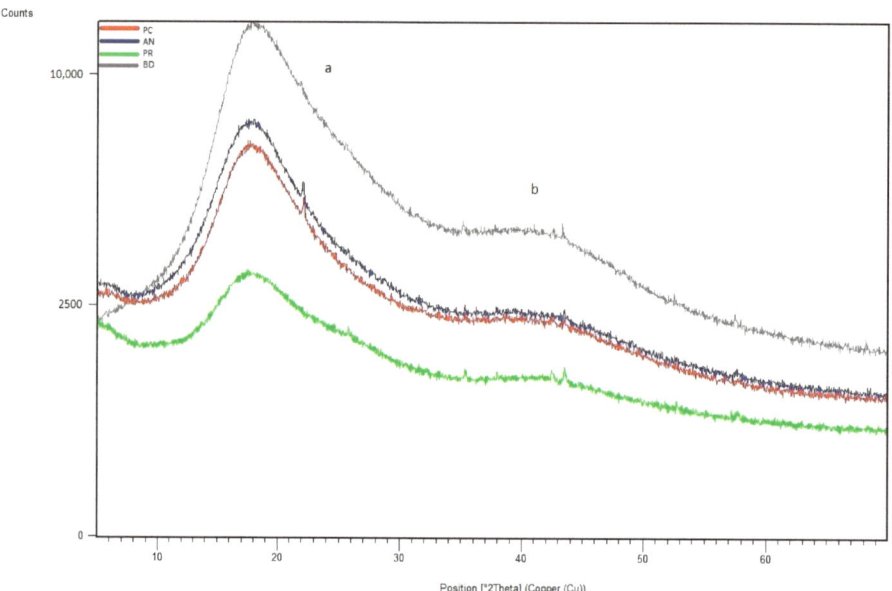

Figure 18. X-ray diffraction results, where different phases were identified as indicated (levels (a,b)).

4. Discussion

The results of the present study showed significant differences ($p < 0.05$) in the concentration of calcium release and pH changes at the studied times. According to the obtained results, the null hypothesis was rejected, since significant differences in the calcium release and pH values were detected between groups.

When the study sample is considered, a previous study by Irawan et al. [18] used 15 teeth per study group, while Cavenago et al, used 10 teeth [20]; in this sense, the present study was consistent with them. The analyses at 3, 24, 72, and 168 h were compared with the related literature, such as the study by Cavenago et al. We extended the analysis to 336 h, 672 h, and 1008 h to obtain comparative data among the various similar studies.

Evaluating the comparisons of the samples' calcium ion release (mg/g), it was considered that the amount of material in the duct, despite the standardization of the samples, could be different. However, the present study analyzed the concentrations in mg/L to compare them with the study of Cavenago et al. [20]. Gandolfi et al. and Irawan et al. obtained the results of the release of calcium ions in ppm [17,18].

In the present study, the pH was calibrated at 4, 7, and 9, instead of at 4, 7, and 14 as previously reported by Cavenago et al.; in this way, the calibrations were easier to perform, since this method contributes to less-dispersed data. The pH data varied little over time and remained constant in all of the included groups. This fact could be due to the fact that the solution used was a buffered medium [23].

Other factors to consider that were not studied here could include the setting time of each material, along with its porosity, since these contribute to its subsequent consistency [24]. Another factor to consider is the oral environment. It should be noted that saliva lipids determine the buffering and antibacterial capacity of the environment, which could influence the pH variations occurring in the medium of calcium silicate cements [25].

According to Palczewska-Komsa et al. [26] the main components of MTA are $CaWO_4$, Ca_3SiO_5, and Ca_2SiO_4 as the main phases of the composition. The content of calcium aluminate improves the biological response of HP MTA. In several scientific studies, the chemical composition of HP MTA was determined using energy-dispersive X-rays. Jiménez-Sánchez et al. [27] found that the structure of the HP MTA tricalcium silicate particles ensures a very close contact between the calcium silicate and calcium aluminate and, thus, favors the hydration reaction. However, according to our records, no W was found. This seems to correlate with the findings of Rochas et al. and Ertas et al., where no W was found in their composition studies of these materials [28,29].

We analyzed the release of calcium ions over time in the groups studied. In our first data record, at 3 h, the release of calcium ions was already observed, consistent with the findings of Gandolfi et al. However, according to the study of Gandolfi et al., it was not until 5 h when apatite formation began. This formation would reach a uniform thickness after 7 days [18].

Significant differences ($p < 0.05$) were observed in the comparisons at 3 h between the AN and PC groups, with the highest value being in the PC group. At 24 h, the PC group showed significant differences again, that time with respect to the BD group. Likewise, there were significant differences between the AN and BD groups, with the AN group being the one that presented a greater release of calcium ions at that point.

At 72 h, there were significant differences ($p < 0.05$) between all groups. This situation was only repeated at 42 days with respect to the AN group. At 28 days (672 h), there were significant differences, where the PR group presented higher figures than the AN group. At 1008 h, the BD group showed significant differences with respect to the AN group. We could see how the behavior of the different cements varied over time, not being comparable between any of the groups. If we compared the variations of our study (in mg/L) to those found in the study of Cavenago et al., we could not establish similarities. It is possible that this was due to the different environments in which the samples were studied [20].

We analyzed the pH values in comparison to similar studies. The pH values increased until reaching 7.93 in the Biodentine group at 7 days. Nevertheless, in the other three groups, the pH value increased up to 24 h and then decreased. The obtained values were consistent with those described by Cavenago et al., although the medium was ultrapure water. The values in the referenced studies could have been higher due to the included medium, in which the samples were not buffered and, therefore, the variation of the ions affected the pH to a greater extent [23].

The results of the study by Ceci et al. showed MTA to have greater pH values than Biodentine. Moreover, the values were higher in their study. The highest values were found for ProRoot MTA, reaching 12.48 at 3 h and 11.56 at 24 h. For all materials tested, a nonsignificant reduction in pH value was recorded after 24 h. [30]. In our study, the same situation seemed to arise for all of the MTA groups after 24 h. The pH results of Ceci et al. were consistent with those of Herrera-Trinidad, who carried out a similar study in an unbuffered medium [31], with ultrapure or deionized water. The pH values in our current study were never below 7.05. However, the pH values could have been lower due to the buffered medium. The buffered solution, according to the distributor, has a pH of 7.2 ± 0.1 at 25 °C. The pH is determined by the amount of H^+ or OH^- ions that is needed until the pH value of a solution changes by 1. Therefore, a high release of H^+/OH^- ions is needed to modify the pH in the medium. According to the study by Irawan et al., bioceramics release more calcium ions and reach higher pH levels than MTA and can increase their tissue regeneration capacity [18]. Our study obtained similar results, with the pH values in Biodentine™ being overall higher than in the other groups. However, there were significant differences in pH at 168 h between Biodentine™ and the MTA groups.

According to Kim et al., due to this alkaline pH, calcium-silicate-based materials can be bactericidal, like calcium hydroxide; nevertheless, they can also induce necrosis of the root surface cells [21]. Another study by De Deus et al., using calcium silicate cements, presented an initial cytotoxic effect, which may have been due to the pH reached on the surface of the cements, which caused the denaturation of adjacent cells and medium proteins [32,33]. This may be related to the slight decreases in pH in the medium observed in the different studies.

Biodentine™ and MTAs have been reported to increase the release of TGF-β1, a modulator of tissue repair and mineralization [34,35], as well as the activation of kinases involved in odontoblastic differentiation [36,37] and cytokines that promote the inflammatory response [38,39]. The study by Machado et al. concurs on the cell proliferation capacity of calcium silicate cements, which, in this case, was significantly greater than in samples with composite [40]. Studies such as that by Benneti et al. showed biocompatibility and mineralization capacity in vivo [41].

According to Rajasekharan et al., each clinical application requires an adequate volume of calcium silicate cement at the repair site, depending on the severity. Furthermore, the area of biomaterial exposed to oral tissues varies widely. Despite the various studies analyzing pH and calcium release, there is no evidence of the environmental effects of pH, volume (Vol), and exposed area on the release of calcium and hydroxyl ions [42].

In addition, it is important to remember that the concentrations found in the solutions do not represent those that reach the enamel or dentin lesion, which would require more complex microanalytical techniques [43]. The beginning of calcium ion release and its implications for apatite formation are important, since, according to Sarkar et al., the biocompatibility, sealing capacity, and dentinogenic activity of MTA stem from the physicochemical reactions between MTA and tissues during hydroxyapatite formation [44].

Hard tissue can form next to MTA due to its alkalinity, sealing ability, and bioactivity [45]. These properties depend on the calcium ion release and pH. These variables were analyzed in our study. It is important to understand how different materials interact with our dental tissues and cells. As reported by Kim et al. [46], stem cells from human exfoliated deciduous teeth and human dental pulp stem cells had significantly low ALP activity after exposure to each material compared with the controls (i.e., cells cultured with osteogenic media). An elevated level of calcium release was found in all calcium silicate cements.

Further studies on natural teeth or in vivo would be necessary to assess variables related to the tooth and/or the environment.

5. Conclusions

Considering the limitations of the present in vitro study, the following conclusions can be drawn:

- There were significant differences in the concentration of calcium release between the different groups studied, with no group being predominant.
- There were no significant differences in the pH values at the timepoints studied, except for the values at 168 h. The pH values showed small variations throughout the evaluated period due to the buffered medium in which the samples were found.

Author Contributions: Conceptualization, R.H.-T., M.L.M. and V.V.-G.; methodology, R.H.-T., M.L.M. and V.V.-G.; validation, R.H.-T., P.M.-M. and V.V.-G.; formal analysis, R.H.-T., P.M.-M. and V.V.-G.; M.L.M.; investigation, R.H.-T.; resources, M.L.M. and V.V.-G.; data curation, R.H.-T. and P.M.-M.; writing—original draft preparation, R.H.-T., P.M.-M., M.F., A.R.W. and V.V.; writing—review and editing, R.H.-T., P.M.-M., M.F., A.R.W. and V.V.; visualization, P.M.-M., M.F., A.R.W. and V.V.; supervision, P.M.-M. and V.V.-G.; project administration, V.V.-G. All authors have read and agreed to the published version of the manuscript.

Funding: This research received no external funding.

Institutional Review Board Statement: Not applicable.

Informed Consent Statement: Not applicable.

Data Availability Statement: The data that support the findings of this study are partially available upon request from the corresponding author.

Conflicts of Interest: The authors declare no conflict of interest.

References

1. Camilleri, J. Investigation of Biodentine as dentine replacement material. *J. Dent.* **2013**, *41*, 600–610. [CrossRef] [PubMed]
2. Witte, D.R. The filling of a root canal with Portland cement. *J. Cent. Assoc. Ger. Dent.* **1878**, *18*, 153–154.
3. Torabinejad, M.; Watson, T.; Ford, T.P. Sealing ability of a mineral trioxide aggregate when used as a root end filling material. *J. Endod.* **1993**, *19*, 591–595. [CrossRef]
4. Torabinejad, M.; White, D.J. Tooth Filling Material and Method. U.S. Patent 5769638, 16 May 1995.
5. Schwartz, R.S.; Mauger, M.; Clement, D.J.; Walker, W.A., 3rd. Mineral trioxide aggregate: A new material for endodontics. *J. Am. Dent. Assoc.* **1999**, *130*, 967–975. [CrossRef]
6. Torabinejad, M.; Chivian, N. Clinical applications of mineral trioxide aggregate. *J. Endod.* **1999**, *25*, 197–205. [CrossRef]
7. Parirokh, M.; Torabinejad, M. Mineral trioxide aggregate: A comprehensive literature review—Part III: Clinical applications, drawbacks, and mechanism of action. *J. Endod.* **2010**, *36*, 400–413. [CrossRef] [PubMed]
8. Margunato, S.; Taşlı, P.N.; Aydın, S.; Kazandağ, M.K.; Şahin, F. In Vitro Evaluation of ProRoot MTA, Biodentine, and MM-MTA on Human Alveolar Bone Marrow Stem Cells in Terms of Biocompatibility and Mineralization. *J. Endod.* **2015**, *41*, 1646–1652. [CrossRef]
9. Antonijevic, D.; Jeschke, A.; Colovic, B.; Milovanovic, P.; Jevremovic, D.; Kisic, D.; Scheidt, A.V.; Hahn, M.; Amling, M.; Jokanovic, V.; et al. Addition of a Fluoride-containing Radiopacifier Improves Micromechanical and Biological Characteristics of Modified Calcium Silicate Cements. *J. Endod.* **2015**, *41*, 2050–2057. [CrossRef]
10. Parirokh, M.; Torabinejad, M. Mineral trioxide aggregate: A comprehensive literature review—Part I: Chemical, physical, and antibacterial properties. *J. Endod.* **2010**, *36*, 16–27. [CrossRef]
11. Gandolfi, M.G.; Siboni, F.; Primus, C.M.; Prati, C. Ion release, porosity, solubility, and bioactivity of mta plus tricalcium silicate. *J. Endod.* **2014**, *40*, 1632–1637. [CrossRef]
12. Giraud, T.; Jeanneau, C.; Bergmann, M.; Laurent, P.; About, I. Tricalcium Silicate Capping Materials Modulate Pulp Healing and Inflammatory Activity In Vitro. *J. Endod.* **2018**, *44*, 1686–1691. [CrossRef] [PubMed]
13. Hakki, S.S.; Bozkurt, B.S.; Ozcopur, B.; Gandolfi, M.G.; Prati, C.; Belli, S. The response of cementoblasts to calcium phosphate resin-based and calcium silicate-based commercial sealers. *Int. Endod. J.* **2013**, *46*, 242–252. [CrossRef] [PubMed]
14. Mehrvarzfar, P.; Abbott, P.V.; Mashhadiabbas, F.; Vatanpour, M.; Savadkouhi, S.T. Clinical and histological responses of human dental pulp to MTA and combined MTA/treated dentin matrix in partial pulpotomy. *Aust. Endod. J.* **2018**, *44*, 46–53. [CrossRef]
15. Wattanapakkavong, K.; Srisuwan, T.; Dent, G.D.C. Release of Transforming Growth Factor Beta 1 from Human Tooth Dentin after Application of Either ProRoot MTA or Biodentine as a Coronal Barrier. *J. Endod.* **2019**, *45*, 701–705. [CrossRef] [PubMed]
16. Costa, F.; Gomes, P.S.; Fernandes, M.H. Osteogenic and Angiogenic Response to Calcium Silicate–based Endodontic Sealers. *J. Endod.* **2016**, *42*, 113–119. [CrossRef] [PubMed]
17. Gandolfi, M.G.; Taddei, P.; Tinti, A.; Prati, C. Apatite-forming ability (bioactivity) of ProRoot MTA. *Int. Endod. J.* **2010**, *43*, 917–929. [CrossRef]
18. Irawan, R.M.; Margono, A.; Djauhari, N. The comparison of calcium ion release and pH changes from modified MTA and bioceramics in regeneration. *J. Phys. Conf. Ser.* **2017**, *884*, 012110. [CrossRef]
19. Natale, L.C.; Rodrigues, M.C.; Xavier, T.A.; Simões, A.; de Souza, D.N.; Braga, R.R. Ion release and mechanical properties of calcium silicate and calcium hydroxide materials used for pulp capping. *Int. Endod. J.* **2015**, *48*, 89–94. [CrossRef]
20. Cavenago, B.C.; Pereira, T.C.; Duarte, M.A.H.; Ordinola-Zapata, R.; Marciano, M.A.; Bramante, C.M.; Bernardineli, N. Influence of powder-to-water ratio on radiopacity, setting time, pH, calcium ion release and a micro-CT volumetric solubility of white mineral trioxide aggregate. *Int. Endod. J.* **2014**, *47*, 120–126. [CrossRef]
21. Kim, M.; Yang, W.; Kim, H.; Ko, H. Comparison of the biological properties of ProRoot MTA, OrthoMTA, and Endocem MTA cements. *J. Endod.* **2014**, *40*, 1649–1653. [CrossRef]

22. Bernardi, A.; Bortoluzzi, E.A.; Felippe, W.T.; Felippe, M.C.S.; Wan, W.S.; Teixeira, C.S. Effects of the addition of nanoparticulate calcium carbonate on setting time, dimensional change, compressive strength, solubility and pH of MTA. *Int. Endod. J.* **2016**, *50*, 97–105. [CrossRef] [PubMed]
23. Urbansky, E.T.; Schock, M.R. Understanding, Deriving, and Computing Buffer Capacity. *J. Chem. Educ.* **2000**, *77*, 1640. [CrossRef]
24. Zakaria, M.N.; Cahyanto, A.; El-Ghannam, A. Calcium release and physical properties of modified carbonate apatite cement as pulp capping agent in dental application. *Biomater. Res.* **2018**, *22*, 35. [CrossRef] [PubMed]
25. Matczuk, J.; Żendzian-Piotrowska, M.; Maciejczyk, M.; Kurek, K. Salivary lipids: A review. *Adv. Clin. Exp. Med.* **2017**, *26*, 1023–1031. [CrossRef] [PubMed]
26. Palczewska-Komsa, M.; Kaczor-Wiankowska, K.; Nowicka, A. New Bioactive Calcium Silicate Cement Mineral Trioxide Aggregate Repair High Plasticity (MTA HP)—A Systematic Review. *Materials* **2021**, *14*, 4573. [CrossRef] [PubMed]
27. Jiménez-Sánchez, M.C.; Segura-Egea, J.J.; Díaz-Cuenca, A. A Microstructure Insight of MTA Repair HP of Rapid Setting Capacity and Bioactive Response. *Materials* **2020**, *13*, 1641. [CrossRef]
28. Rocha, A.C.R.; Padrón, G.H.; Garduño, M.V.G.; Aranda, R.L.G. Physicochemical analysis of MTA Angelus®and Biodentine®conducted with X ray difraction, dispersive energy spectrometry, X ray fluorescence, scanning electron microscope and infra red spectroscopy. *Rev. Odontológica Mex.* **2015**, *19*, e170–e176. [CrossRef]
29. Ertas, H.; Kucukyilmaz, E.; Ok, E.; Uysal, B. Push-out bond strength of different mineral trioxide aggregates. *Eur. J. Dent.* **2014**, *8*, 348–352. [CrossRef]
30. Poggio, C.; Ceci, M.; Beltrami, R.; Chiesa, M.; Colombo, M. Biological and chemical-physical properties of root-end filling materials: A comparative study. *J. Conserv. Dent.* **2015**, *18*, 94–99. [CrossRef]
31. Herrera Trinidad, R.; García Barbero, E. Liberación de iones calcio y ph de los cementos de silicato de calcio. 2019. Available online: https://eprints.ucm.es/57362/ (accessed on 3 July 2023).
32. De Deus, G.; Ximenes, R.; Gurgel-Filho, E.D.; Plotkowski, M.C.; Coutinho-Filho, T. Cytotoxicity of MTA and Portland cement on human ECV 304 endothelial cells. *Int. Endod. J.* **2005**, *38*, 604–609. [CrossRef]
33. Min, K.-S.; Park, H.-J.; Lee, S.-K.; Park, S.-H.; Hong, C.-U.; Kim, H.-W.; Lee, H.-H.; Kim, E.-C. Effect of Mineral Trioxide Aggregate on Dentin Bridge Formation and Expression of Dentin Sialoprotein and Heme Oxygenase-1 in Human Dental Pulp. *J. Endod.* **2008**, *34*, 666–670. [PubMed]
34. Giraud, T.; Jeanneau, C.; Rombouts, C.; Bakhtiar, H.; Laurent, P.; About, I. Pulp capping materials modulate the balance between inflammation and regeneration. *Dent. Mater.* **2019**, *35*, 24–35. [CrossRef] [PubMed]
35. Laurent, P.; Camps, J.; About, I. Biodentine™ induces TGF-β1 release from human pulp cells and early dental pulp mineralization. *Int. Endod. J.* **2012**, *45*, 439–448. [PubMed]
36. Chen, I.; Salhab, I.; Setzer, F.C.; Kim, S.; Nah, H.-D. A New Calcium Silicate–based Bioceramic Material Promotes Human Osteo- and Odontogenic Stem Cell Proliferation and Survival via the Extracellular Signal-regulated Kinase Signaling Pathway. *J. Endod.* **2016**, *42*, 480–486. [CrossRef]
37. Jung, J.-Y.; Woo, S.-M.; Lee, B.-N.; Koh, J.-T.; Nör, J.E.; Hwang, Y.-C. Effect of Biodentine and Bioaggregate on odontoblastic differentiation via mitogen-activated protein kinase pathway in human dental pulp cells. *Int. Endod. J.* **2015**, *48*, 177–184. [CrossRef] [PubMed]
38. Chang, S.W.; Bae, W.J.; Yi, J.K.; Lee, S.; Lee, D.W.; Kum, K.Y.; Kim, E.C. Odontoblastic Differentiation, Inflammatory Response, and Angiogenic Potential of 4 Calcium Silicate–Based Cements: Micromega MTA, ProRoot MTA, RetroMTA, and Experimental calcium Silicate Cement. *J. Endod.* **2015**, *41*, 1524–1529. [CrossRef]
39. Chang, S.-W.; Lee, S.-Y.; Kum, K.-Y.; Kim, E.-C. Effects of ProRoot MTA, Bioaggregate, and Micromega MTA on Odontoblastic Differentiation in Human Dental Pulp Cells. *J. Endod.* **2014**, *40*, 113–118. [CrossRef]
40. Machado, J.; Johnson, J.D.; Paranjpe, A. The Effects of Endosequence Root Repair Material on Differentiation of Dental Pulp Cells. *J. Endod.* **2016**, *42*, 101–105. [CrossRef]
41. Benetti, F.; Gomes-Filho, J.E.; de Araújo Lopes, J.M.; Barbosa, J.G.; Jacinto, R.C.; Cintra, L.T. In vivo biocompatibility and biomineralization of calcium silicate cements. *Eur. J. Oral Sci.* **2018**, *126*, 326–333.
42. Rajasekharan, S.; Vercruysse, C.; Martens, L.; Verbeeck, R. Effect of Exposed Surface Area, Volume and Environmental pH on the Calcium Ion Release of Three Commercially Available Tricalcium Silicate Based Dental Cements. *Materials* **2018**, *11*, 123, Correction *Materials* **2021**, *14*, 340. [CrossRef]
43. Braga, R.R.; About, I. How far do calcium release measurements properly reflect its multiple roles in dental tissue mineralization? *Clin. Oral Investig.* **2019**, *23*, 501. [CrossRef] [PubMed]
44. Sarkar, N.; Caicedo, R.; Ritwik, P.; Moiseyeva, R.; Kawashima, I. Physicochemical Basis of the Biologic Properties of Mineral Trioxide Aggregate. *J. Endod.* **2005**, *31*, 97–100. [CrossRef] [PubMed]
45. AlShwaimi, E.; Majeed, A.; Ali, A.A. Pulpal Responses to Direct Capping with Betamethasone/Gentamicin Cream and Mineral Trioxide Aggregate: Histologic and Micro–Computed Tomography Assessments. *J. Endod.* **2016**, *42*, 30–35. [CrossRef] [PubMed]
46. Kim, B.; Lee, Y.-H.; Kim, I.-H.; Lee, K.E.; Kang, C.-M.; Lee, H.-S.; Choi, H.-J.; Cheon, K.; Song, J.S.; Shin, Y. Biocompatibility and mineralization potential of new calcium silicate cements. *J. Dent. Sci.* **2023**, *18*, 1189–1198. [CrossRef] [PubMed]

Disclaimer/Publisher's Note: The statements, opinions and data contained in all publications are solely those of the individual author(s) and contributor(s) and not of MDPI and/or the editor(s). MDPI and/or the editor(s) disclaim responsibility for any injury to people or property resulting from any ideas, methods, instructions or products referred to in the content.

Article

The Combined Effects on Human Dental Pulp Stem Cells of Fast-Set or Premixed Hydraulic Calcium Silicate Cements and Secretome Regarding Biocompatibility and Osteogenic Differentiation

Yun-Jae Ha [1], Donghee Lee [2] and Sin-Young Kim [1,*]

1. Department of Conservative Dentistry, Seoul St. Mary's Hospital, College of Medicine, The Catholic University of Korea, Seoul 06591, Republic of Korea; gkdbswo1122@naver.com
2. Department of Dentistry, College of Medicine, The Catholic University of Korea, Seoul 06591, Republic of Korea; dong524@catholic.ac.kr
* Correspondence: jeui99@catholic.ac.kr; Tel.: +82-2-2258-1787

Citation: Ha, Y.-J.; Lee, D.; Kim, S.-Y. The Combined Effects on Human Dental Pulp Stem Cells of Fast-Set or Premixed Hydraulic Calcium Silicate Cements and Secretome Regarding Biocompatibility and Osteogenic Differentiation. *Materials* **2024**, *17*, 305. https://doi.org/10.3390/ma17020305

Academic Editors: Luigi Generali, Vittorio Checchi and Eugenio Pedullà

Received: 9 December 2023
Revised: 1 January 2024
Accepted: 4 January 2024
Published: 7 January 2024

Copyright: © 2024 by the authors. Licensee MDPI, Basel, Switzerland. This article is an open access article distributed under the terms and conditions of the Creative Commons Attribution (CC BY) license (https:// creativecommons.org/licenses/by/ 4.0/).

Abstract: An important part of regenerative endodontic procedures involving immature permanent teeth is the regeneration of the pulp–dentin complex with continuous root development. Hydraulic calcium silicate cements (HCSCs) are introduced for the pulpal treatment of immature permanent teeth. The stem-cell-derived secretome recently has been applied for the treatment of various damaged tissues. Here, we evaluated the biocompatibility and osteogenic differentiation of HCSCs combined with secretome on human dental pulp stem cells. In the Cell Counting Kit-8 test and wound healing assays, significantly higher cell viability was observed with secretome application. In alkaline phosphatase analysis, the activity was significantly higher with secretome application in all groups, except for RetroMTA on day 2 and Endocem MTA Premixed on day 4. In an Alizarin Red S staining analysis, all groups with secretome application had significantly higher staining values. Quantitative real-time polymerase chain reaction results showed that the day 7 expression of *OSX* significantly increased with secretome application in all groups. *SMAD1* and *DSPP* expression also increased significantly with secretome addition in all groups except for Biodentine. In conclusion, HCSCs showed favorable biocompatibility and osteogenic ability and are predicted to demonstrate greater synergy with the addition of secretome during regenerative endodontic procedures involving immature permanent teeth.

Keywords: cell migration; cell viability; hydraulic calcium silicate cement; osteogenic ability; secretome

1. Introduction

With the development of tissue engineering, the regeneration and replacement of new tissues using stem cells, growth factors, and scaffolds have been studied in the field of regenerative medicine. Within endodontics, regenerative endodontics is emerging as an area of focus for addressing necrotic pulp tissue. The aim of a regenerative endodontic procedure is to replace damaged tooth structures, including the root, dentin, and cells within the pulp–dentin complex [1]. Rehabilitation of the pulp–dentin complex takes on greater significance when immature permanent teeth suffer from necrosis.

Mineral trioxide aggregate (MTA) is applied extensively in pulp therapy for immature permanent teeth, offering exceptional biocompatibility, antibacterial properties, and an impressive sealing capability [2]. MTA is a biomaterial and has been used in various endodontic treatments, such as vital pulp therapy, regenerative endodontic procedures, repair of perforation sites, and retrograde filling during endodontic microsurgery [3–7]. The classic MTA, ProRoot MTA (Dentsply Tulsa Dental Specialties, Tulsa, OK, USA), is renowned for its capacity to enhance osteogenic and dentinogenic markers as well as promote angiogenesis in human dental pulp stem cells (hDPSCs) [3]. Its disadvantages are related to its long setting time, the likelihood of discoloration, and its heavy metal content [4–7].

To compensate for these shortcomings, various MTAs and hydraulic calcium silicate cements (HCSCs) have been developed [8,9]. HCSCs were intended to shorten the setting time and overcome discoloration [10–12]. RetroMTA (BioMTA, Seoul, Republic of Korea) has biocompatibility and physical properties that are similar to those of ProRoot MTA, and its initial setting time is faster than ProRoot MTA, at 150 s [13,14]. Among HCSCs, in animal and clinical trials, Biodentine (Septodont, Saint-Maur-dens-Fossés, France) has shown bone-formation and dentin-formation capabilities similar to those of ProRoot MTA, without cytotoxicity, and its initial setting time is also faster than ProRoot MTA with 12 min [15,16]. Additionally, a newly developed material, Endocem MTA Premixed (Maruchi, Wonju, Republic of Korea), is an injectable premixed material that offers technicians a simplified clinical option. [17,18]

The secretome derived from human dental stem cells, including stem cells derived from human exfoliated deciduous teeth (SHEDs), has emerged as a promising candidate for addressing various medical conditions. It shows potential in mitigating neurodegenerative diseases, repairing nerve damage, treating cartilage defects, and promoting bone regeneration through mechanisms that include nerve protection, anti-inflammatory effects, cell death inhibition, and the stimulation of angiogenesis [19]. As indicated recently, the secretome derived from hDPSCs demonstrates a versatile applicability in the field of tissue regeneration. This versatility is attributed to its multifaceted paracrine effects, and notably, it exhibits a remarkable capacity for osteogenesis [19,20].

The objective of this study was to evaluate and compare the biocompatibility and osteogenic differentiation capacity of ProRoot MTA, Biodentine, RetroMTA, and Endocem MTA Premixed when used in combination with secretome on hDPSCs.

2. Materials and Methods

2.1. Human Dental Pulp Stem Cells (hDPSCs)

This study received approval from the Institutional Review Board of Seoul St. Mary's Hospital, the Catholic University of Korea (IRB No. MC22ZASI0063). We employed hDPSCs that are commercially available from Top Cell Bio Inc. (Seoul, Republic of Korea) (SH0010, Passage 4). The company's evaluation confirmed that CD29, CD44, CD73, and CD105 expression was >90% positive in all samples of these cell lines.

Cells were cultivated in a growth medium, which included HyClone minimum essential medium (α-MEM; Cytiva, Marlborough, MA, USA), supplemented with 10% fetal bovine serum (Cytiva), penicillin, and streptomycin. To assess osteogenic potential, we used an osteogenic medium consisting of α-MEM along with beta-glycerophosphate (Sigma-Aldrich, St. Louis, MO, USA), dexamethasone (Sigma-Aldrich), and ascorbic acid (Sigma-Aldrich). Cell cultures were incubated in a controlled environment at 37 °C under 5% CO_2, with rigorous adherence to sterile conditions throughout all experimental procedures.

2.2. Disks of Different MTAs and HCSCs for Experimentation

ProRoot MTA and HCSCs (RetroMTA, Biodentine, Endocem MTA Premixed), and the SHEDs-derived secretome (Top Cell Bio Inc.) were used in this study. Table 1 lists the composition details for each of them. With strict adherence to sterile protocols, we prepared each cement following the manufacturer's guidelines and produced specimens in the form of disks measuring 6 mm in diameter and 3 mm in height. ProRoot MTA powder was mixed with the supplied liquid in a ratio of 3:1, one capsule of Biodentine powder was mixed with five drops of supplied liquid using a mixer, one cap of RetroMTA powder was mixed with three drops of supplied liquid, and Endocem MTA Premixed was directly applied to the mold. Subsequently, all samples were enveloped in wet gauze and left to harden for 72 h within a room-temperature clean bench. Afterward, they underwent a 4 h sterilization process using ultraviolet rays within a clean bench.

Table 1. Materials: manufacturers and chemical compositions.

Material	Manufacturer	Composition	Powder Size	Batch Number
ProRoot MTA	Dentsply Tulsa Dental Specialties, Tulsa, OK, USA	Portland cement (tricalcium silicate, dicalcium silicate, and tricalcium aluminate) 75% Calcium sulfate dihydrate (gypsum) 5% Bismuth oxide 20%	6.9 μm	294002
Biodentine	Septodont, Saint-Maur-dens-Fossés, France	Tricalcium silicate 80.1% Calcium carbonate 14.9% Zirconium Oxide 5% Calcium chloride and polycarboxylate as an aqueous liquid	3.77 μm	B29557
RetroMTA	BioMTA, Seoul, Republic of Korea	Calcium carbonate 60–80% Silicon dioxide 5–15% Aluminum oxide 5–10% Calcium zirconia complex 20–30%	2.62 μm	RMCA04D03
Endocem MTA Premixed	Maruchi, Wonju, Republic of Korea	Zirconium dioxide 45–55% Calcium silicate 20–25% Calcium aluminate 1–5% Calcium sulfate 1–5% Dimethyl sulfoxide 20–25% Thickening agent 1–5%		FD220323A
secretome	Top Cell Bio Inc., Seoul, Republic of Korea			

2.3. Categorization of Experimental Groups

Experimental groups were developed as follows: a control group without secretome, consisting of hDPSCs cultured without experimental disks; a control group with secretome, consisting of hDPSCs cultured without experimental disks, and with culture media supplemented only with secretome; ProRoot MTA without secretome; ProRoot MTA with secretome; Biodentine without secretome; Biodentine with secretome; RetroMTA without secretome; RetroMTA with secretome; Endocem MTA Premixed without secretome; and Endocem MTA Premixed with secretome.

2.4. Cell Viability Assay

The cytotoxic effects of MTAs and HCSCs were evaluated using a Cell Counting Kit-8 test (CCK-8) (CK04-13; Dojindo, Kumamoto, Japan). The cell proliferation rate of hDPSCs was determined at different timepoints, including immediately after culturing (0), as well as at 2, 4, and 6 days post-culture. hDPSCs were plated into 24-well cell culture plates (SPL Life Sciences, Pocheon, Republic of Korea) at a density of 2.5×10^4 cells per well in a growth medium. After 24 h of cell adhesion culture, absorbance immediately after culture was measured.

Each disk was positioned on a cell co-culture platform (SPLInsert; SPL Life Sciences) featuring 0.4 μm pores, and this platform was positioned directly above the hDPSCs in each well. To ensure that the culture solution remained at the full height of the disk, we supplemented each well with an extra 1 mL of culture solution. Control groups consisted of hDPSCs cultured without experimental disks. We introduced 20 μL of CCK-8 solution into each well, followed by incubation at 37 °C for 1 h. We then measured the absorbance at 450 nm, using 650 nm as the reference point, with the aid of an absorbance microplate reader

(PowerWave XS; BioTek Instruments, Winooski, VT, USA). Eight independent samples were assessed for each group.

2.5. Cell Migration Assay

We assessed cell migration ability using a wound healing assay. hDPSCs were cultured in a 24-well cell culture plate at a density of 3.5×10^4 cells per well in a growth medium. Following a 24 h incubation period, using a 1000 µL pipette tip, we created a scratch in the center of the cell layer adhered to the well floor. After the scratching, isolated cell substances were washed with phosphate-buffered saline. After 24 h of incubation, each disk was placed on a SPLInsert with pores of 0.4 µm, and the platform was placed above hDPSCs in each well. To maintain the culture solution to the full height of the disk, each well was supplemented with an additional 1 mL of culture solution. hDPSCs with experimental disks were cultured for 4 days, and we obtained wound healing images at 0, 1, 2, 3, and 4 days using a phase-contrast microscope from Olympus (Tokyo, Japan). The area covered by cell migration towards the scratch site was quantified using ImageJ 1.46r software (National Institutes of Health, Bethesda, MD, USA). Cell migration into the scratch area was then calculated as a percentage relative to the initial scratch size. Four independent samples were assessed for each group.

2.6. Alkaline Phosphatase (ALP) Activity

To evaluate the osteogenic potential of hDPSCs, we conducted ALP analysis at 2 and 4 days. Cells were cultivated at a density of 0.7×10^4 cells per well in a 24-well cell culture plate containing osteogenic medium. Individual disks were positioned on an SPLInsert with 0.4 µm pores, situated above the hDPSCs in each well. To maintain the culture solution to the full height of the disk, we supplemented each well with an additional 1 mL of culture solution. On days 2 and 4, each well was rinsed with phosphate-buffered saline, followed by the addition of 20 µL of a dissolution buffer (0.2% Triton X-100; AnaSpec, Fremont, CA, USA). The blend was incubated at 37 °C for 15 min. Subsequently, we detached the adherent cells by scraping, transferred the solution to a 1.5 mL microcentrifuge tube, and centrifuged it at 4 °C for 10 min at $2500 \times g$. The resulting supernatant was collected for ALP analysis. For the ALP analysis, 50 µL of p-nitrophenyl phosphate (pNPP) ALP substrate solution (AnaSpec, Fremont, CA, USA) was added to each well. pNPP was placed on top of each sample, and the reagent was gently mixed for 30 s. After incubation at 4 °C for 30 min, absorbance was measured at 405 nm with a PowerWave XS. Six independent samples were assessed for each group.

2.7. Alizarin Red S (ARS) Staining Assay

We used the ARS staining assay to assess calcium deposition within hDPSCs. Each experimental disk was combined with an osteogenic medium and stored in a 100% humidity incubator at 37 °C for 6 days, resulting in a concentration of 5 mg/mL. hDPSCs were cultured in a 24-well plate at a density of 0.7×10^4 cells per well and exposed to an experimental disk exudate solution for 12 days. Subsequently, we fixed the cells using a 4% paraformaldehyde solution and a 2% ARS solution from ScienCell (Carlsbad, CA, USA) for 20 min. The dyeing process involved treatment with 10% cetylpyridinium chloride (Sigma-Aldrich) for 15 min, followed by absorbance measurement at 560 nm using a PowerWave XS. Six independent samples were assessed for each group.

2.8. Quantitative Real-Time Polymerase Chain Reaction (qRT-PCR)

Seven days after culture, we obtained total RNA from each cell with the RNeasy Mini Kit (Qiagen, Hilden, Germany). Subsequently, we initiated RNA reverse transcription reactions using the RevertAid First Strand cDNA Synthesis Kit (Thermo Fisher Scientific, Waltham, MA, USA) to synthesize cDNA. For the analysis, we employed primers for key markers, including Runt-related transcription factor 2 (*RUNX2*), osterix (*OSX*), and Suppressor of Mothers against Decapentaplegic (*SMAD1*), which are known osteogenic

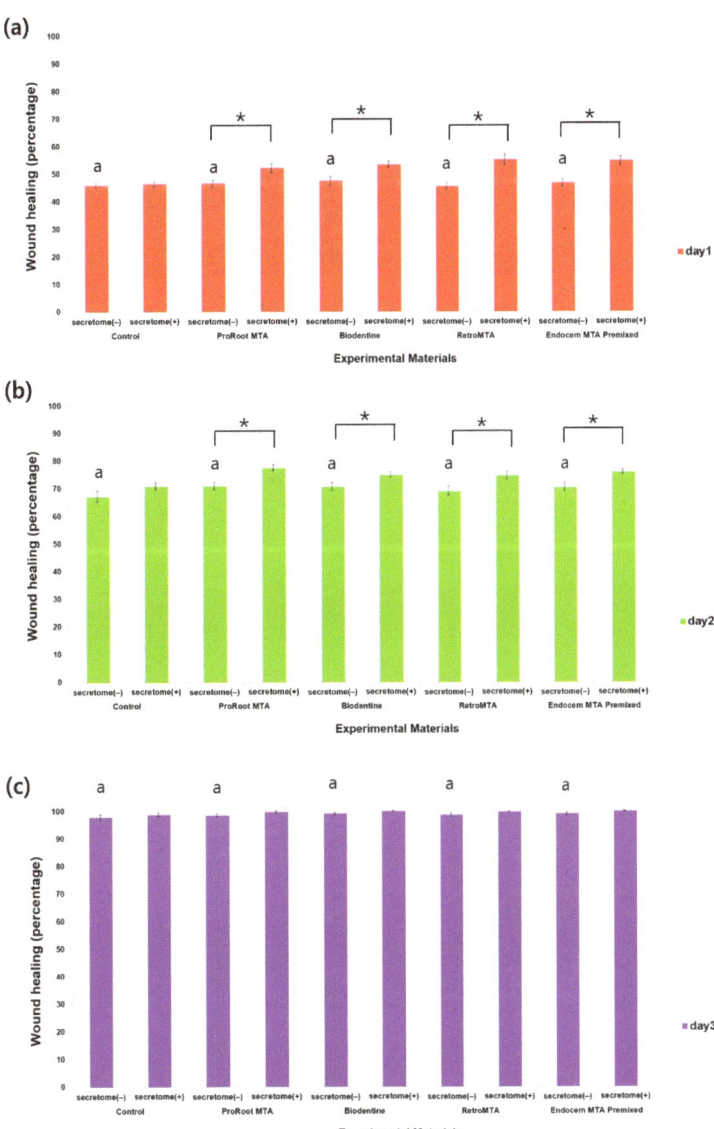

Figure 2. Comparison of the cell migration values for each material with or without secretome on (**a**) day 1, (**b**) day 2, and (**c**) day 3 of the wound healing assay. Different characters indicate statistical significance between compared groups, and * denotes statistical significance within the indicated comparison.

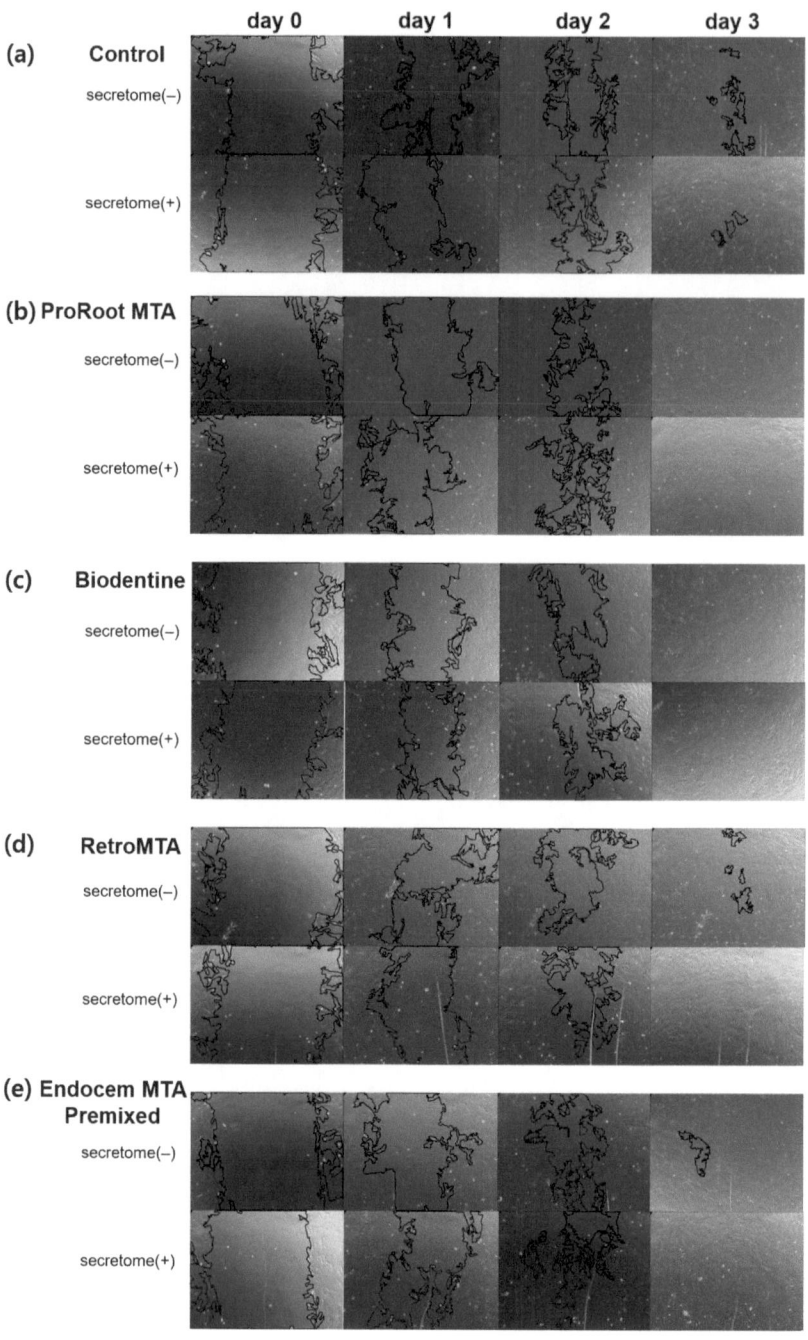

Figure 3. Comparison of representative images from wound healing assay. (**a**) The control group, (**b**) the ProRoot MTA group, (**c**) the Biodentine group, (**d**) the RetroMTA group, (**e**) the Endocem MTA Premixed group.

3.3. ALP Activity

ALP activity was measured on days 2 and 4, and higher ALP activity was observed over time (Figure 4). On day 2, ALP activity was significantly higher in all groups with secretome application compared to the corresponding groups without secretome, except for the RetroMTA group (Figure 4a, $p < 0.05$). On day 4, ALP activity was significantly higher in all groups with versus without secretome except for the Endocem MTA Premixed group (Figure 4b, $p < 0.05$). On day 2, the results were significantly higher in the RetroMTA group without added secretome compared with the other groups (Figure 4a, $p < 0.05$).

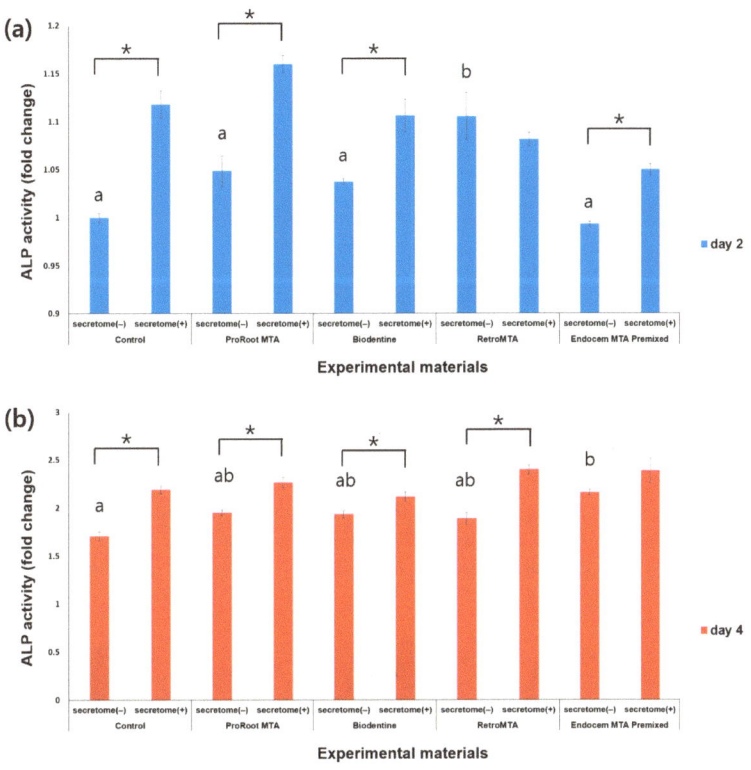

Figure 4. Comparison of ALP activity for each material according to the presence or absence of secretome on (**a**) day 2 and (**b**) day 4. Different characters indicate statistical significance between compared groups, and * denotes statistical significance within the indicated comparison.

3.4. ARS Staining Assay

ARS staining was performed on days 6 and 12 after incubation, and ARS staining levels were observed to increase over time (Figure 5). On day 6, the ARS staining level in the RetroMTA group was higher than in other groups with or without secretome. On day 6, for all experimental groups and the control group, values were significantly higher with versus without secretome (Figure 5a, $p < 0.05$). On day 12, values were significantly higher in all experimental groups than in the control group, regardless of whether secretome was applied (Figure 5b, $p < 0.05$), and with versus without secretome in all groups, including the control group (Figure 5b, $p < 0.05$). The ARS staining images for each group are shown in Figure 6. From day 6 to day 12, the area of calcium accumulation increased, and, with secretome application, the ARS staining area increased.

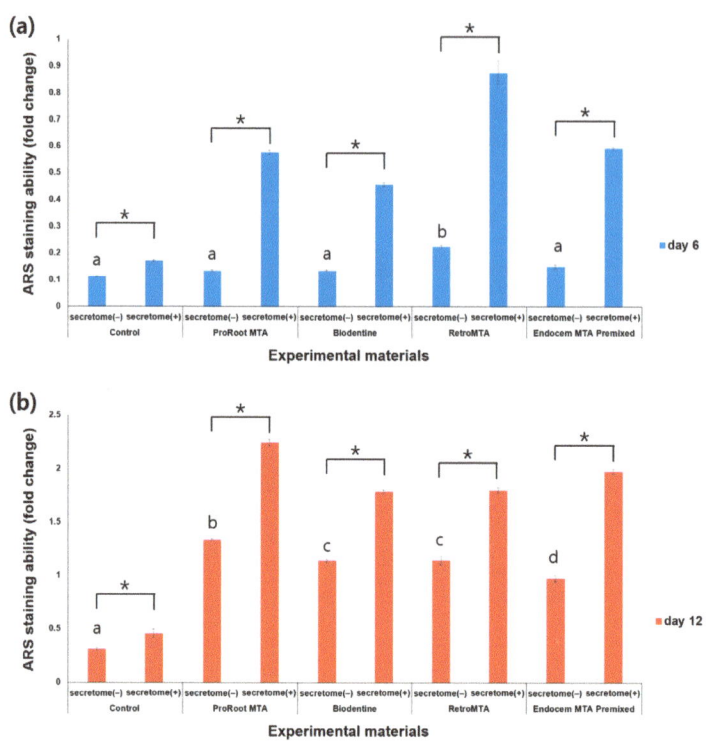

Figure 5. Comparison of ARS staining for each material with or without secretome on (**a**) day 6 and (**b**) day 12. Different characters indicate statistical significance between compared groups, and * denotes statistical significance within the indicated comparison.

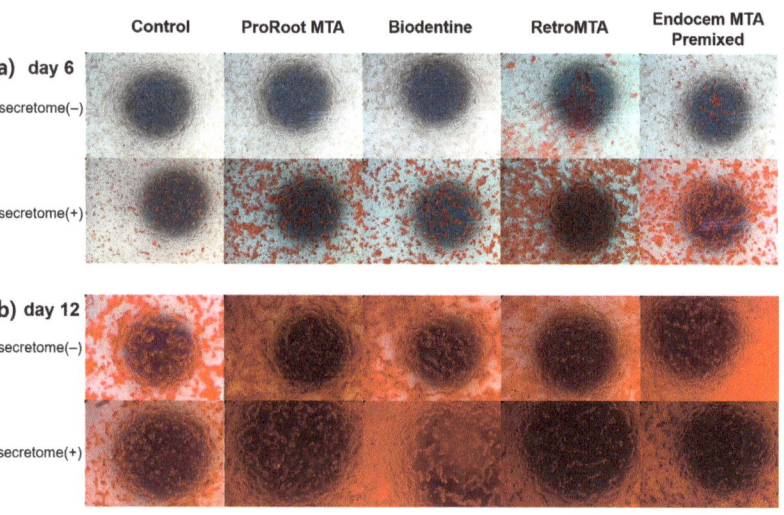

Figure 6. Representative images from the ARS staining assay on (**a**) day 6 and (**b**) day 12.

3.5. qRT-PCR

With the application of secretome, the expression level of all markers was generally increased (Figure 7). This increase was especially prominent for *OSX* in all experimental groups with secretome compared to their respective groups without it (Figure 7b, $p < 0.05$). In the case of *SMAD1*, the Endocem MTA Premixed group showed increased expression compared to other groups regardless of secretome application (Figure 7c, $p < 0.05$). Both *SMAD1* and *DSPP* expression levels increased with secretome in all but the Biodentine group compared to the respective groups without secretome (Figure 7c,d, $p < 0.05$).

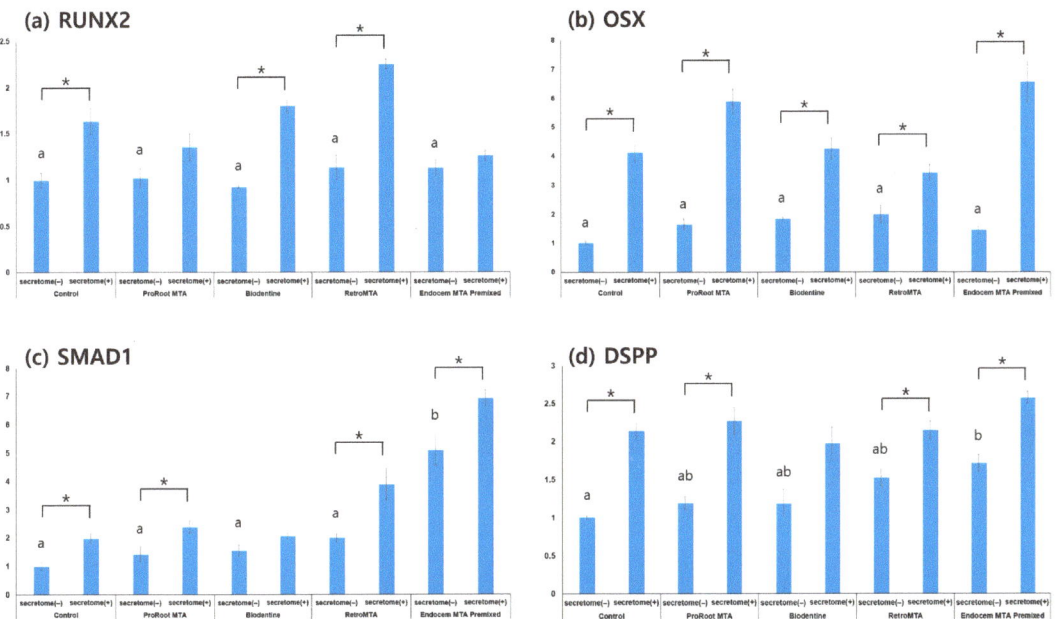

Figure 7. Comparison of qRT-PCR results in each experimental group on day 7. (**a**) *RUNX2*, (**b**) *OSX*, (**c**) *SMAD-1*, (**d**) *DSPP*. Different characters indicate statistical significance between compared groups, and * denotes statistical significance within the indicated comparison.

4. Discussion

When the pulp of an immature permanent tooth with a wide root apex shows necrosis, calcium hydroxide has historically been used to form hard tissue at the root apex. However, if calcium hydroxide is used inside the root canal for a long period, the root canal wall can weaken, leading to root fracture, with the disadvantage of this requiring a long treatment period [1]. To compensate, MTA, with its excellent sealing power, biocompatibility, and antibacterial properties, was introduced for the pulp therapy of immature permanent teeth [2]. However, because a regenerative endodontic procedure for immature permanent teeth is carried out in a blood environment, and the hardening of the material can be reduced, HCSCs have been developed to improve on the disadvantages of existing MTAs, such their as different operability and long hardening time [21,22].

Recently, a secretome derived from hDPSCs has been applied in the treatment of various damaged tissues and has shown significant osteogenic potential because of its various paracrine effects [19]. Mesenchymal stem cells derived from hDPSCs are emerging as a promising tool, with great potential for tooth and oral tissue regeneration, and have been used in clinical trials as a tool in regenerative medicine [23].

In this study, given the osteogenic potential and tissue regeneration effects of secretome, we sought to determine if its use in addition to MTA or HCSCs could increase the success rate of vital pulp therapy or regenerative endodontic procedures. For this purpose, we assessed the biocompatibility and osteogenic potential of ProRoot MTA, Biodentine, RetroMTA, and Endocem MTA Premixed, each with added secretome.

The key finding of this study was that all experimental hydraulic calcium silicate cements showed favorable biocompatibility like ProRoot MTA, and significantly higher cell viability was observed with secretome application. Regarding osteogenic potential, all experimental hydraulic calcium silicate cements showed higher mineralization activity when secretome was combined, as did ProRoot MTA.

To assess biocompatibility, we used CCK-8 analysis and the wound healing assay. CCK-8 relies on Dojindo's water-soluble tetrazolium salt (WST-8) in a sensitive colorimetric assay to determine viable cell counts. WST-8 is reduced by cellular dehydrogenase to produce formazan, which dissolves in tissue culture medium, and CCK-8 leverages the fact that the amount of formazan dye is directly proportional to the number of living cells [24]. The methyl thiazol tetrazolium (MTT) assay also is widely used to evaluate cell proliferation and cytotoxicity [25] but has the disadvantage of underestimating cell damage and detecting cell death only in the late stages when the cell metabolism is significantly reduced [26]. Therefore, in this study, we used the CCK-8 assay, which has superior sensitivity and low cytotoxicity compared with the MTT assay and allows for repeated measurements of the same sample. The wound healing assay is used to evaluate biocompatibility by observing cell migration to the scratch site [27].

CCK-8 and wound healing assay results showed an overall trend of cell proliferation over time. Based on CCK-8 assay findings, cells proliferated better when secretome was applied, even on day 6. In the wound healing assay, however, secretome had an effect only in the early stages, and cell migration was not faster when secretome was added from day 3. MTA and HCSCs did not have a negative effect on cell survival and migration and showed a higher cell viability than the control group on day 2 based on the CCK-8 assay. A previously published study also showed that when MTA was applied to SHEDs, cell viability was similar to or higher than that of the control group [28]. In addition, other studies have shown that tricalcium silicate, an ingredient in ProRoot MTA and Biodentine, induces the proliferation of human dental pulp cells [29], and that RetroMTA, which contains calcium carbonate, yielded better cell viability compared with a control group [30]. These previous results are consistent with the current findings. Moreover, the additional application of secretome appears to further increase biocompatibility, similar to previous results showing that the conditioned medium of hDPSCs significantly increases cell differentiation and migration [31].

In regenerative endodontic procedures on immature permanent teeth, the increased root wall is reported to be formed by the growth of cementum, bone, or dentin-like tissue, based on histological observation. In a study using teeth extracted after a regenerative endodontic procedure, newly formed mineralized tissue along the dentin wall was cementum-like tissue or osteoid dentin [32]. In the current work, secretome was mixed with the expectation that it would affect regeneration rather than repair when applied with MTA and HCSCs to hDPSCs. We used ALP activity, ARS analysis, and qRT-PCR to evaluate the osteogenic and dentinogenic potential of these combinations.

ALP initially attaches to the outside of the cell membrane, including in osteoblasts, and is used as a marker to distinguish cells. When active bone formation increases, the expression of ALP increases, which means that osteoblast differentiation actively occurs [33]. We found significantly higher ALP activity in the MTA and HCSC groups compared with the control group at day 4, and when secretome was applied during the entire experimental period, ALP activity significantly increased in the ProRoot MTA and Biodentine groups. Thus, MTA or HCSCs support hDPSC's differentiation into osteoblasts, and the effect appears to be greater when secretome is applied. Silva et al. reported that MTA exhibits a high pH, which can neutralize acids secreted by osteoclasts and help prevent the further destruction of mineralized tissue [34]. In one study, ProRoot MTA in interaction with water generated calcium hydroxide and calcium ion release, leading to a high pH and mineralization, and consequently increased ALP activation [35]. In addition, previous studies have reported that Biodentine, RetroMTA, and Endocem MTA Premixed increase

ALP activation [17,18], which supports the current finding that the experimental groups treated with HCSCs on day 4 had higher ALP values than the control group.

The ARS staining assay is used to determine the degree of osteogenic differentiation by evaluating calcium accumulation in cell culture. Here, we observed that ARS staining values increased on days 6 and 12 in all groups with secretome compared to the respective groups without secretome. On day 12, ARS staining values for the experimental groups with MTA or HCSCs were higher than the control values, regardless of whether secretome was applied. MTA and HCSCs thus increase calcium accumulation in cells, an effect that is further increased when secretome is added. Previous studies have shown that secretome derived from human gingiva demonstrates osteogenic potential in the form of significant calcium deposits, even after 6 weeks [20]. This finding is similar to the current results, in which calcium nodule formation continued to amplify for up to 12 days when secretome was added.

In this study, qRT-PCR was used to analyze the expression levels of the osteogenic markers *RUNX2*, *OSX*, and *SMAD1* and the dentinogenic marker *DSPP* in each experimental group. *RUNX2* and *OSX* are early markers of osteoblast differentiation [36,37], and *RUNX2* is expressed in predontoblasts, is essential for tooth formation, and is involved in the mineralization of tooth tissue [38,39]. SMAD1 is a protein involved in enhancing the differentiation of hDPSCs into odontoblasts and osteoblasts [40]. DMP-1 is a non-collagenous protein that is important for the mineralization of both bone and tooth [41]. There are reports that *DMP-1* is observed in abundance in the early stage of odontoblast differentiation, and *DSPP* is reported to be observed in the predentin layer but was also found in fully polarized dentin in the late stage of differentiation [42]. In this study, when secretome was not applied, overall, there was no statistically significant difference at 7 days in *RUNX2* and *OSX* expression levels between the control group and all experimental groups. Thus, it is not possible to confirm from these findings that adding MTA or HCSCs amplified the expression of osteogenic and dentinogenic genes compared to the control group. However, when secretome was applied, there were generally significantly highly expressed gene levels in the control group, as well as in all experimental groups. In the case of the ProRoot MTA, RetroMTA, and Endocem MTA Premixed groups, *OSX*, *SMAD1*, and *DSPP* showed higher expression levels when secretome was added. Applying secretome, therefore, can enhance osteogenic and dentinogenic cell differentiation of hDPSCs, and adding secretome to MTA or HCSCs can further increase this effect. In a previous study, ProRoot MTA application led to the increased expression of *osteocalcin (OCN)*, *SMAD1*, *OSX*, *DMP-1*, and *DSPP* compared with controls, and the expression of *OCN*, *SMAD1*, *OSX*, and *DSPP* was amplified with the application of Endocem MTA Premixed [18]. Another study showed that *RUNX2* expression increased with the addition of Biodentine [43,44], consistent with the current results.

The main purpose of pulp therapy for immature permanent teeth is to preserve the activity and function of the remaining pulp tissue and to increase root wall thickness and root length. Accordingly, the materials used for the pulp therapy of immature permanent teeth must induce the differentiation of hDPSCs into osteogenic and dentinogenic cells to enable the formation of the pulp–dentin complex and reparative dentin [45]. Here, we found that MTA and HCSCs increased osteogenic potential, and when secretome was added, both osteogenic and dentinogenic potential improved. In this way, using MTA and HCSCs together with secretome in the pulp therapy of immature permanent teeth could more effectively achieve root wall thickness and root length growth.

Recently, increasing evidence indicates the importance of paracrine signaling induced by mesenchymal stem cells as a support mechanism for the regeneration of damaged tissues [46,47]. Although some studies have shown that the differentiation capacity of mesenchymal stem cells is not the primary mechanism for repairing damaged tissue in most diseases, tissue repair properties may be attributed to bioactive factors secreted by these stem cells that contribute to their paracrine activity [46–49]. Cell-free treatment using the secretome derived from mesenchymal stem cells could represent a new approach in regenerative medicine [46,47,49–51]. Secretome derived from hDPSCs and SHEDs may contribute to promoting new bone formation, tooth tissue regeneration, and nerve regeneration through the secretion of proangiogenic factors [19,20], and promote bone formation by enhancing the expression of osteogenic genes including *RUNX2*, *OCN*, *osteopontin*, and *OSX* [19], in keeping with the current results. In vitro studies have shown that secretome derived from DPSCs can have a stimulating effect on odontoblast differentiation, and in vivo studies point to its induction of the regeneration of pulp-like tissue [20].

This study had several limitations. It was conducted on cells formed as a monolayer in vitro, and in vivo research is needed. The problem of allogenic immune responses in major histocompatibility complex-mismatched recipients reported in previous studies must be considered when applying the secretome in vivo. In addition to this experiment, further methods to evaluate physical and chemical properties (e.g., solubility, volume stability, pH, Vickers hardness test) should be implemented to compare the differences and superiority between materials. There could be potential discrepancies between the material composition provided by the manufacturer and its actual surface composition. Despite these limitations, a conclusion can be drawn based on the current findings. The various HCSCs developed to overcome the shortcomings of existing MTAs in vital pulp therapy or regenerative endodontic procedures on immature permanent teeth show good biocompatibility and osteogenic potential, in agreement with other available evidence. If secretome is added, a more significant synergistic effect on cell viability and osteogenic potential can be expected.

Author Contributions: Y.-J.H., D.L. and S.-Y.K. designed the experimental method. Y.-J.H. and D.L. performed all experimental procedures and obtained all experimental data. Y.-J.H. and D.L. analyzed all experimental results and Y.-J.H. and S.-Y.K. interpreted them. All authors have read and agreed to the published version of the manuscript.

Funding: This research was supported by a National Research Foundation of Korea (NRF) grant funded by the Ministry of Science, ICT, and Future Planning (NRF-2019R1F1A1058955) and Ministry of Education (NRF-2021R1I1A2041534).

Institutional Review Board Statement: This research was approved by the Catholic University of Korea (IRB No. MC22ZASI0063).

Informed Consent Statement: Not applicable.

Data Availability Statement: The datasets used/or analyzed during the current study are available from the corresponding author on reasonable request.

Conflicts of Interest: The authors declare that there are no conflicts of interest regarding this study. The funders had no role related to the results of this study.

References

1. Kim, S.G.; Malek, M.; Sigurdsson, A.; Lin, L.M.; Kahler, B. Regenerative Endodontics: A Comprehensive Review. *Int. Endod. J.* **2018**, *51*, 1367–1388. [CrossRef] [PubMed]
2. Parirokh, M.; Torabinejad, M. Mineral Trioxide Aggregate: A Comprehensive Literature Review--Part I: Chemical, Physical, and Antibacterial Properties. *J. Endod.* **2010**, *36*, 16–27. [CrossRef] [PubMed]
3. Youssef, A.R.; Emara, R.; Taher, M.M.; Al-Allaf, F.A.; Almalki, M.; Almasri, M.A.; Siddiqui, S.S. Effects of Mineral Trioxide Aggregate, Calcium Hydroxide, Biodentine and Emdogain on Osteogenesis, Odontogenesis, Angiogenesis and Cell Viability of Dental Pulp Stem Cells. *BMC Oral Health* **2019**, *19*, 133. [CrossRef] [PubMed]
4. Ber, B.S.; Hatton, J.F.; Stewart, G.P. Chemical Modification of ProRoot MTA to Improve Handling Characteristics and Decrease Setting Time. *J. Endod.* **2007**, *33*, 1231–1234. [CrossRef]
5. Boutsioukis, C.; Noula, G.; Lambrianidis, T. Ex Vivo Study of the Efficiency of Two Techniques for the Removal of Mineral Trioxide Aggregate Used as a Root Canal Filling Material. *J. Endod.* **2008**, *34*, 1239–1242. [CrossRef]
6. Dominguez, M.S.; Witherspoon, D.E.; Gutmann, J.L.; Opperman, L.A. Histological and Scanning Electron Microscopy Assessment of Various Vital Pulp-Therapy Materials. *J. Endod.* **2003**, *29*, 324–333. [CrossRef]
7. Roberts, H.W.; Toth, J.M.; Berzins, D.W.; Charlton, D.G. Mineral Trioxide Aggregate Material Use in Endodontic Treatment: A Review of the Literature. *Dent. Mater.* **2008**, *24*, 149–164. [CrossRef]

8. Torabinejad, M.; Hong, C.U.; McDonald, F.; Pitt Ford, T.R. Physical and Chemical Properties of a New Root-End Filling Material. *J. Endod.* **1995**, *21*, 349–353. [CrossRef]
9. Dawood, A.E.; Parashos, P.; Wong, R.H.K.; Reynolds, E.C.; Manton, D.J. Calcium Silicate-Based Cements: Composition, Properties, and Clinical Applications. *J. Investig. Clin. Dent.* **2017**, *8*, e12195. [CrossRef]
10. Malkondu, Ö.; Karapinar Kazandağ, M.; Kazazoğlu, E.A. Review on Biodentine, a Contemporary Dentine Replacement and Repair Material. *Biomed. Res. Int.* **2014**, *2014*, 160951. [CrossRef]
11. Ma, J.; Shen, Y.; Stojicic, S.; Haapasalo, M. Biocompatibility of Two Novel Root Repair Materials. *J. Endod.* **2011**, *37*, 793–798. [CrossRef] [PubMed]
12. Sarkar, N.K.; Caicedo, R.; Ritwik, P.; Moiseyeva, R.; Kawashima, I. Physicochemical Basis of the Biologic Properties of Mineral Trioxide Aggregate. *J. Endod.* **2005**, *31*, 97–100. [CrossRef] [PubMed]
13. Chung, C.J.; Kim, E.; Song, M.; Park, J.W.; Shin, S.J. Effects of Two Fast-Setting Calcium-Silicate Cements on Cell Viability and Angiogenic Factor Release in Human Pulp-Derived Cells. *Odontology* **2016**, *104*, 143–151. [CrossRef] [PubMed]
14. Wongwatanasanti, N.; Jantarat, J.; Sritanaudomchai, H.; Hargreaves, K.M. Effect of Bioceramic Materials on Proliferation and Odontoblast Differentiation of Human Stem Cells from the Apical Papilla. *J. Endod.* **2018**, *44*, 1270–1275. [CrossRef]
15. Nowicka, A.; Lipski, M.; Parafiniuk, M.; Sporniak-Tutak, K.; Lichota, D.; Kosierkiewicz, A.; Kaczmarek, W.; Buczkowska-Radlińska, J. Response of Human Dental Pulp Capped with Biodentine and Mineral Trioxide Aggregate. *J. Endod.* **2013**, *39*, 743–747. [CrossRef] [PubMed]
16. Shin, M.; Chen, J.W.; Tsai, C.Y.; Aprecio, R.; Zhang, W.; Yochim, J.M.; Teng, N.; Torabinejad, M. Cytotoxicity and Antimicrobial Effects of a New Fast-Set MTA. *Biomed. Res. Int.* **2017**, *2017*, 2071247. [CrossRef] [PubMed]
17. Kim, H.M.; Lee, D.; Kim, S.Y. Biocompatibility and Osteogenic Potential of Calcium Silicate-Based Cement Combined with Enamel Matrix Derivative: Effects on Human Bone Marrow-Derived Stem Cells. *Materials* **2021**, *14*, 7750. [CrossRef]
18. Kim, Y.; Lee, D.; Kye, M.; Ha, Y.J.; Kim, S.Y. Biocompatible Properties and Mineralization Potential of Premixed Calcium Silicate-Based Cements and Fast-Set Calcium Silicate-Based Cements on Human Bone Marrow-Derived Mesenchymal Stem Cells. *Materials* **2022**, *15*, 7595. [CrossRef]
19. Bar, J.K.; Lis-Nawara, A.; Grelewski, P.G. Dental Pulp Stem Cell-Derived Secretome and Its Regenerative Potential. *Int. J. Mol. Sci.* **2021**, *22*, 12018. [CrossRef]
20. El Moshy, S.; Radwan, I.A.; Rady, D.; Abbass, M.M.S.; El-Rashidy, A.A.; Sadek, K.M.; Dörfer, C.E.; Fawzy El-Sayed, K.M. Dental Stem Cell-Derived Secretome/Conditioned Medium: The Future for Regenerative Therapeutic Applications. *Stem Cells Int.* **2020**, *2020*, 7593402. [CrossRef]
21. Çelik, B.N.; Mutluay, M.S.; Arikan, V.; Sari, Ş. The Evaluation of MTA and Biodentine as a Pulpotomy Materials for Carious Exposures in Primary Teeth. *Clin. Oral Investig.* **2019**, *23*, 661–666. [CrossRef] [PubMed]
22. Vafaei, A.; Nikookhesal, M.; Erfanparast, L.; Løvschall, H.; Ranjkesh, B. Vital Pulp Therapy Following Pulpotomy in Immature First Permanent Molars with Deep Caries Using Novel Fast-Setting Calcium Silicate Cement: A Retrospective Clinical Study. *J. Dent.* **2022**, *116*, 103890. [CrossRef] [PubMed]
23. Grawish, M.E.; Saeed, M.A.; Sultan, N.; Scheven, B.A. Therapeutic Applications of Dental Pulp Stem Cells in Regenerating Dental, Periodontal and Oral-Related Structures. *World J. Meta Anal.* **2021**, *9*, 176–192. [CrossRef]
24. Xia, S.; Liu, M.; Wang, C.; Xu, W.; Lan, Q.; Feng, S.; Qi, F.; Bao, L.; Du, L.; Liu, S.; et al. Inhibition of SARS-CoV-2 (Previously 2019-nCoV) Infection by a Highly Potent Pan-Coronavirus Fusion Inhibitor Targeting Its Spike Protein That Harbors a High Capacity to Mediate Membrane Fusion. *Cell Res.* **2020**, *30*, 343–355. [CrossRef] [PubMed]
25. Gorduysus, M.; Avcu, N.; Gorduysus, O.; Pekel, A.; Baran, Y.; Avcu, F.; Ural, A.U. Cytotoxic Effects of Four Different Endodontic Materials in Human Periodontal Ligament Fibroblasts. *J. Endod.* **2007**, *33*, 1450–1454. [CrossRef]
26. Wei, W.; Qi, Y.P.; Nikonov, S.Y.; Niu, L.N.; Messer, R.L.W.; Mao, J.; Primus, C.M.; Pashley, D.H.; Tay, F.R. Effects of an Experimental Calcium Aluminosilicate Cement on the Viability of Murine Odontoblast-Like Cells. *J. Endod.* **2012**, *38*, 936–942. [CrossRef]
27. Luo, J.; Xu, J.; Cai, J.; Wang, L.; Sun, Q.; Yang, P. The In Vitro and In Vivo Osteogenic Capability of the Extraction Socket-Derived Early Healing Tissue. *J. Periodontol.* **2016**, *87*, 1057–1066. [CrossRef]
28. Araújo, L.B.; Cosme-Silva, L.; Fernandes, A.P.; Oliveira, T.M.; Cavalcanti, B.D.N.; Gomes Filho, J.E.; Sakai, V.T. Effects of Mineral Trioxide Aggregate, BiodentineTM and Calcium Hydroxide on Viability, Proliferation, Migration and Differentiation of Stem Cells from Human Exfoliated Deciduous Teeth. *J. Appl. Oral Sci.* **2018**, *26*, e20160629. [CrossRef]
29. Du, R.; Wu, T.; Liu, W.; Li, L.; Jiang, L.; Peng, W.; Chang, J.; Zhu, Y. Role of the Extracellular Signal-Regulated Kinase 1/2 Pathway in Driving Tricalcium Silicate-Induced Proliferation and Biomineralization of Human Dental Pulp Cells In Vitro. *J. Endod.* **2013**, *39*, 1023–1029. [CrossRef]
30. Souza, L.C.; Yadlapati, M.; Dorn, S.O.; Silva, R.; Letra, A. Analysis of Radiopacity, pH and Cytotoxicity of a New Bioceramic Material. *J. Appl. Oral Sci.* **2015**, *23*, 383–389. [CrossRef]
31. Kawamura, R.; Hayashi, Y.; Murakami, H.; Nakashima, M. EDTA Soluble Chemical Components and the Conditioned Medium from Mobilized Dental Pulp Stem Cells Contain an Inductive Microenvironment, Promoting Cell Proliferation, Migration, and Odontoblastic Differentiation. *Stem Cell Res. Ther.* **2016**, *7*, 77. [CrossRef] [PubMed]
32. Simon, S.R.; Tomson, P.L.; Berdal, A. Regenerative Endodontics: Regeneration or Repair? *J. Endod.* **2014**, *40*, S70–S75. [CrossRef] [PubMed]

33. Sharma, U.; Pal, D.; Prasad, R. Alkaline Phosphatase: An Overview. *Indian J. Clin. Biochem.* **2014**, *29*, 269–278. [CrossRef] [PubMed]
34. Silva, E.J.; Rosa, T.P.; Herrera, D.R.; Jacinto, R.C.; Gomes, B.P.; Zaia, A.A. Evaluation of Cytotoxicity and Physicochemical Properties of Calcium Silicate-Based Endodontic Sealer MTA Fillapex. *J. Endod.* **2013**, *39*, 274–277. [CrossRef] [PubMed]
35. Ali, M.R.W.; Mustafa, M.; Bårdsen, A.; Bletsa, A. Tricalcium Silicate Cements: Osteogenic and Angiogenic Responses of Human Bone Marrow Stem Cells. *Eur. J. Oral Sci.* **2019**, *127*, 261–268. [CrossRef] [PubMed]
36. Wu, J.; Li, N.; Fan, Y.; Wang, Y.; Gu, Y.; Li, Z.; Pan, Y.; Romila, G.; Zhou, Z.; Yu, J. The Conditioned Medium of Calcined Tooth Powder Promotes the Osteogenic and Odontogenic Differentiation of Human Dental Pulp Stem Cells via MAPK Signaling Pathways. *Stem Cells Int.* **2019**, *2019*, 4793518. [CrossRef] [PubMed]
37. Wei, X.; Li, J.; Liu, H.; Niu, C.; Chen, D. Salidroside Promotes the Osteogenic and Odontogenic Differentiation of Human Dental Pulp Stem Cells Through the BMP Signaling Pathway. *Exp. Ther. Med.* **2022**, *23*, 55. [CrossRef]
38. Camilleri, S.; McDonald, F. Runx2 and Dental Development. *Eur. J. Oral Sci.* **2006**, *114*, 361–373. [CrossRef]
39. Miyazaki, T.; Kanatani, N.; Rokutanda, S.; Yoshida, C.; Toyosawa, S.; Nakamura, R.; Takada, S.; Komori, T. Inhibition of the Terminal Differentiation of Odontoblasts and Their Transdifferentiation into Osteoblasts in Runx2 Transgenic Mice. *Arch. Histol. Cytol.* **2008**, *71*, 131–146. [CrossRef]
40. Shang, W.; Xiong, S. Phenytoin Is Promoting the Differentiation of Dental Pulp Stem Cells into the Direction of Odontogenesis/Osteogenesis by Activating BMP4/Smad Pathway. *Dis. Markers* **2022**, *2022*, 7286645. [CrossRef]
41. Jacob, A.; Zhang, Y.; George, A. Transcriptional Regulation of Dentin Matrix Protein 1 (DMP1) in Odontoblasts and Osteoblasts. *Connect. Tissue Res.* **2014**, *55*, 107–112. [CrossRef] [PubMed]
42. Suzuki, S.; Haruyama, N.; Nishimura, F.; Kulkarni, A.B. Dentin Sialophosphoprotein and Dentin Matrix Protein-1: Two Highly Phosphorylated Proteins in Mineralized Tissues. *Arch. Oral Biol.* **2012**, *57*, 1165–1175. [CrossRef] [PubMed]
43. Chung, M.; Lee, S.; Chen, D.; Kim, U.; Kim, Y.; Kim, S.; Kim, E. Effects of Different Calcium Silicate Cements on the Inflammatory Response and Odontogenic Differentiation of Lipopolysaccharide-Stimulated Human Dental Pulp Stem Cells. *Materials* **2019**, *12*, 1295. [CrossRef] [PubMed]
44. Daltoé, M.O.; Paula-Silva, F.W.; Faccioli, L.H.; Gatón-Hernández, P.M.; De Rossi, A.; Bezerra Silva, L.A. Expression of Mineralization Markers During Pulp Response to Biodentine and Mineral Trioxide Aggregate. *J. Endod.* **2016**, *42*, 596–603. [CrossRef] [PubMed]
45. Zhao, X.; He, W.; Song, Z.; Tong, Z.; Li, S.; Ni, L. Mineral Trioxide Aggregate Promotes Odontoblastic Differentiation Via Mitogen-Activated Protein Kinase Pathway in Human Dental Pulp Stem Cells. *Mol. Biol. Rep.* **2012**, *39*, 215–220. [CrossRef] [PubMed]
46. Kumar, P.; Kandoi, S.; Misra, R.; Vijayalakshmi, S.; Rajagopal, K.; Verma, R.S. The Mesenchymal Stem Cell Secretome: A New Paradigm Towards Cell-Free Therapeutic Mode in Regenerative Medicine. *Cytokine Growth Factor Rev.* **2019**, *46*, 1–9. [CrossRef]
47. Harrell, C.R.; Fellabaum, C.; Jovicic, N.; Djonov, V.; Arsenijevic, N.; Volarevic, V. Molecular Mechanisms Responsible for Therapeutic Potential of Mesenchymal Stem Cell-Derived Secretome. *Cells* **2019**, *8*, 467. [CrossRef]
48. Teixeira, F.G.; Salgado, A.J. Mesenchymal Stem Cells Secretome: Current Trends and Future Challenges. *Neural Regen. Res.* **2020**, *15*, 75–77. [CrossRef]
49. Gwam, C.; Mohammed, N.; Ma, X. Stem Cell Secretome, Regeneration, and Clinical Translation: A Narrative Review. *Ann. Transl. Med.* **2021**, *9*, 70. [CrossRef]
50. Teixeira, F.G.; Carvalho, M.M.; Sousa, N.; Salgado, A.J. Mesenchymal Stem Cells Secretome: A New Paradigm for Central Nervous System Regeneration? *Cell. Mol. Life Sci.* **2013**, *70*, 3871–3882. [CrossRef]
51. Sultan, N.; Amin, L.E.; Zaher, A.R.; Scheven, B.A.; Grawish, M.E. Dental Pulp Stem Cells: Novel Cell-Based and Cell-Free Therapy for Peripheral Nerve Repair. *World J. Stomatol.* **2019**, *7*, 1–19. [CrossRef]

Disclaimer/Publisher's Note: The statements, opinions and data contained in all publications are solely those of the individual author(s) and contributor(s) and not of MDPI and/or the editor(s). MDPI and/or the editor(s) disclaim responsibility for any injury to people or property resulting from any ideas, methods, instructions or products referred to in the content.

Article

Antibacterial Activity and Sustained Effectiveness of Calcium Silicate-Based Cement as a Root-End Filling Material against *Enterococcus faecalis*

Seong-Hee Moon [1,2,†], Seong-Jin Shin [2,†], Seunghan Oh [1,2] and Ji-Myung Bae [1,2,*]

1. Institute of Biomaterials & Implant, College of Dentistry, Wonkwang University, 460 Iksan-daero, Iksan City 54538, Republic of Korea; shmoon06@gmail.com (S.-H.M.); shoh@wku.ac.kr (S.O.)
2. Department of Dental Biomaterials, College of Dentistry, Wonkwang University, 460 Iksan-daero, Iksan City 54538, Republic of Korea; ko2742@naver.com
* Correspondence: baejimy@wku.ac.kr
† These authors contributed equally to this work.

Citation: Moon, S.-H.; Shin, S.-J.; Oh, S.; Bae, J.-M. Antibacterial Activity and Sustained Effectiveness of Calcium Silicate-Based Cement as a Root-End Filling Material against *Enterococcus faecalis*. *Materials* 2023, 16, 6124. https://doi.org/10.3390/ma16186124

Academic Editors: Luigi Generali, Vittorio Checchi and Eugenio Pedullà

Received: 24 August 2023
Revised: 5 September 2023
Accepted: 6 September 2023
Published: 8 September 2023

Copyright: © 2023 by the authors. Licensee MDPI, Basel, Switzerland. This article is an open access article distributed under the terms and conditions of the Creative Commons Attribution (CC BY) license (https:// creativecommons.org/licenses/by/ 4.0/).

Abstract: Several calcium silicate cement (CSC) types with improved handling properties have been developed lately for root-end filling applications. While sealing ability is important, a high biocompatibility and antimicrobial effects are critical. This study aimed to conduct a comparative evaluation of the antimicrobial efficacy and sustained antibacterial effectiveness against *Enterococcus faecalis* (*E. faecalis*) of commercially available CSCs mixed with distilled water (DW) and chlorhexidine (CHX). Various products, viz., ProRoot mixed with DW (PRW) or with CHX (PRC), Endocem mixed with DW (EW) or with CHX (EC), and Endocem premixed (EP) syringe type, were used. While antibacterial activity against *E. faecalis* was evaluated using a direct contact method, the specimens were stored in a shaking incubator for 30 d for antibacterial sustainability. The cytotoxicity was evaluated using a cell counting kit-8 assay in human periodontal ligament stem cells. The antibacterial activities of EP, EW, and EC were greater than those of PRC and PRW ($p < 0.05$). The antibacterial sustainability of EP was the highest without cytotoxicity for up to 30 days ($p < 0.05$). In conclusion, the pre-mixed injectable type EP was most effective in terms of antibacterial activity and sustained antibacterial effectiveness without cytotoxicity.

Keywords: calcium silicate cement; *Enterococcus faecalis*; antibacterial; sustained antibacterial effectiveness

1. Introduction

Ever since the development of calcium silicate cement (CSC)—also known as Mineral Trioxide Aggregate (MTA)—by Dr. Torabinejad in the early 1990s, it has been widely used in various applications of endodontics, such as root canal obturation, perforation repair, and root-end filling [1,2]. Although materials such as zinc oxide eugenol, IRM, Cavit, and amalgam were used for root-end filling previously, these materials had drawbacks that included the potential for microleakage and associated bacterial growth [3]. CSC has become the gold standard for root-end filling not only owing to its biocompatibility but also because of its superior antibacterial effects compared to other materials [4,5].

Enterococcus faecalis (*E. faecalis*) is a Gram-positive facultative anaerobic bacterium that is one of the microorganisms most commonly associated with treatment failure during endodontic procedures [6]. *E. faecalis* is resistant to the high pH induced by calcium hydroxide [7]. Therefore, it is imperative that materials used in root-end fillings possess superior antimicrobial activities and sustained antimicrobial properties against *E. faecalis*. Previous studies have reported methods to enhance the antimicrobial properties of CSC [8,9]. One such approach is the incorporation of chlorhexidine (CHX), which is an antimicrobial agent that is widely used in dentistry, into CSC [10–12]. Although few studies have demonstrated the antimicrobial efficacy of CSCs mixed with CHX [11,13], generally there has been a lack of research on the sustained antimicrobial properties of CSCs, particularly against *E. faecalis*.

Even though CSCs are known to exhibit a high biocompatibility and sealing ability, they possess certain inherent limitations such as an extended setting time and difficulty in manipulation [14]. Recently, improved CSC types have been developed to overcome these drawbacks. Especially, ENDOCEM MTA (Maruchi, Wonju, Republic of Korea), which is a quick-setting type CSC with a setting time of 4 min as per the manufacturer's specifications, offers several benefits owing to its rapid setting time. ENDOCEM premixed (Maruchi) was introduced as an alternative injectable-type CSC recently. This type of CSC offers several advantages, including the elimination of mixing requirements, convenient handling, and enhanced applicability. Various studies have been conducted on the physicochemical characteristics of these new CSC types [15–18]. Nevertheless, there is a notable dearth of studies that focus on evaluating the antibacterial activities of fast-set and premixed-type CSC. Especially, their sustained antibacterial effectiveness over 10 days has not been reported yet.

Therefore, this study aimed to compare the antibacterial activity and sustained antibacterial effectiveness for up to 30 days of several CSCs containing fast-set and premixed types. Additionally, the cytotoxicity of periodontal ligament stem cells and radiopacity were also evaluated. The null hypothesis is that there are no significant differences in antibacterial effect, sustained antibacterial effect, and radiopacity among the groups.

2. Materials and Methods

2.1. Materials and Preparation of the Specimens

The various types of MTA cement used in this study are listed in Table 1. One set of specimens was prepared by mixing ProRoot MTA with distilled water (PRW) and another set was prepared by mixing ProRoot MTA with chlorhexidine (PRC; Hexamedine, Bukwang Pharm. Co., Ltd., Ansan, Republic of Korea) at a water-to-powder (W/P) ratio of 0.18 cc/500 mg, according to the manufacturer's instructions. Similarly, Endocem MTA was mixed with distilled water (EW) as well as chlorhexidine (EC) at a W/P ratio of 0.12 cc/300 mg. Because the injectable paste Endocem MTA Premixed Regular (EP) does not require mixing, it was injected directly into a silicone mold. Distilled water is used to mix the powder of CSC powders indicated in the instructions for use and was used in the PRW and EW groups as control groups in comparison with the use of CHX (PRC and EC groups). The specimens were prepared in the form of a disc with 6 mm diameter and 1 ± 0.1 mm thickness using a mold and allowed to set at 37 °C for 24 h in conditions of relative humidity above 90%. Subsequently, the specimens were sterilized by ultraviolet (UV) irradiation on a clean bench (JSCB-1200SB, JSR, Gongju, Republic of Korea) for 30 min. The distance from the UV light was approximately 50 mm.

Table 1. Information on the calcium silicate cements used in this study.

Trade Name	Code	Liquid Type	Powder or Paste Composition	Manufacturer
Pro Root MTA	PRW	Distilled water	Tricalcium silicate, Dicalcium silicate, Bismuth oxide, Tricalcium aluminate, calcium sulfate dehydrate, tetra calcium aluminoferrite, gypsum, calcium oxide	Dentsply, Tulsa, TN, USA
	PRC	Chlorhexidine		
Endocem MTA	EW	Distilled water	Tricalcium silicate, Dicalcium silicate, Bismuth oxide, Tricalcium aluminate	Maruchi, Wonju, Republic of Korea
	EC	Chlorhexidine		
Endocem MTA Premixed Regular	EP	Injectable type paste	Zirconium dioxide, Calcium silicate, calcium aluminate, Calcium sulfate, Dimethyl sulfoxide, Lithium carbonate, Thickening agents	Maruchi, Wonju, Republic of Korea

2.2. Antibacterial Activity

The antibacterial activity of the specimens was evaluated against *E. faecalis* (KCTC) by adopting the direct contact method for the antibacterial test [19]. Bacteria were incubated aerobically in Brain Heart Infusion (BHI; BD DIFCO, Detroit, MI, USA) broth at 37 °C. After setting, each specimen was placed in the well of a 96-well plate ($n = 6$), and then each specimen was inoculated with 10 µL of E. faecalis (approximately 10^7 bacteria); the specimens were then incubated at 37 °C for 1 h (Figure 1). Subsequently, 200 µL of BHI broth was added by pipetting and mixed, and 15 µL of the suspension was transferred to a new well plate filled with 215 µL of fresh BHI broth. After incubation at 37 °C for 6 h, the optical density (OD) values were measured to compare the bacterial growth of mid-log phase using a microplate reader (SpectraMax 250; Molecular Devices Co., San Jose, CA, USA). The OD values for each group were converted into percentages by considering the value of the negative control group as the base.

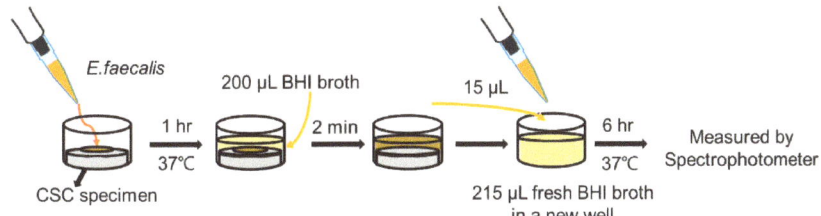

Figure 1. Schematic illustration of the antibacterial test.

2.3. Sustained Antibacterial Effectiveness

To evaluate sustained antibacterial effectiveness, only those groups that exhibited antibacterial activity were chosen for testing. Disc shaped specimens similar to those used in the antibacterial assays were prepared ($n = 6$). Each specimen was placed in a 1.5 mL microtube filled with 1 mL PBS and stored in a shaking incubator at 120 rpm maintained at 37 °C for a predefined time duration of 1, 2, 10, 20, and 30 days. After expiry of the predefined time, the corresponding specimens were shifted to 96-well plates and subjected to the same process against E. faecalis using direct contact methods, as in the case of antibacterial assay. The OD values for each group were converted into percentages by considering the value of the negative control group as the base.

2.4. Cytotoxicity on Human Periodontal Ligament Stem Cell

Disc shaped specimens similar to those used in the antibacterial assays were prepared and eluted in 10 mL of the cell culture media (MEM Alpha, Gibco, Carlsbad, CA, USA) for 72 h at 37 °C according to ISO 10993-12:2012 [20]. Human periodontal ligament stem cells (hPDLSCs; Celprogen, Torrance, CA, USA) were distributed into 96-well plates at a density of 1×10^4 cells/well and incubated for 24 h. The eluted media of each group was filtered through a 0.2 µm syringe filter (SC25P020SS, HYUNDAI MICRO Co., Ltd., Seoul, Republic of Korea) and applied to the cells using a method which is a slight modification of that suggested in ISO 10993-5 [21]. Briefly, hPDLSCs were seeded at a density of 1×10^4 cells/well into a 96-well plate and incubated for 24 h. Subsequently, 100 µL of the extracted media was added to each well and incubated for 1, 2, and 3 days. Cell viability was measured using the Cell Counting Kit-8 (CCK-8; Dojindo Molecular Technologies, Rockville, MD, USA) following the manufacturer's instructions. The OD value was measured at 450 nm using a microplate reader (SpectraMax 250). The negative control (NC) refers to cells that were cultured without any extract media.

2.5. Scanning Electron Microscopic Analysis

The surface morphology of all the specimens, i.e., those after the antibacterial assay as well as after sustained antibacterial effectiveness on the 10th day, were analyzed by field-emission scanning electron microscopy (FE-SEM; S-4800, Hitachi, Japan) at 5 kV under ×5000 and ×10,000 magnifications. Before observation, for fixation of the bacteria on the specimens, 2.5% glutaraldehyde (Sigma-Aldrich, St. Louis, MO, USA) was applied for

4 h. Then, the specimens were dehydrated by immersing them in 60, 70, 80, 90, and 100% ethanol (Sigma-Aldrich) for 10 min each. After drying the remaining ethanol, the specimens were coated using a platinum coater (E-1045; Hitachi Ltd., Tokyo, Japan).

2.6. Radiopacity

The radiopacities of the set materials were measured according to the test methods recommended in ISO 13116:2014 [22]. The specimens were disc shaped with 6 mm diameter and 1 ± 0.1 mm thickness ($n = 6$). To obtain radiographic images, each of the specimens were placed on a digital sensor along with an aluminum step wedge and exposed to an X-ray unit (Carestream Dental, Stuttgart, Germany) at 60 kV and 10 mA with a 300 mm focus-film distance. The gray pixel values of each specimen and the aluminum step wedge were determined using Image J (NIH, Bethesda, MD, USA). A linear equation was drawn using the grayscale of an aluminum stem wedge and the measured grayscale from the specimen was used to calculate the corresponding thickness of Al in millimeters.

2.7. Statistical Analysis

Statistical analyses were performed using IBM SPSS Statistics for Windows (version 26.0, IBM Corp., Armonk, NY, USA). Sustained antibacterial effectiveness data were analyzed by one-way analysis of variance (ANOVA) and Welch's test with Games-Howell multiple comparisons ($\alpha = 0.05$). Other data were analyzed using one-way ANOVA with Tukey's multiple comparisons as a post hoc test ($\alpha = 0.05$).

3. Results

3.1. Antibacterial Activity

The growth of *E. faecalis* was significantly inhibited in EW, EC, EP, and PRC ($p < 0.05$) (Figure 2). In particular, the EW, EC, and EP groups showed higher bacterial inhibition than the PRC group ($p < 0.05$) with no significant difference compared to the positive control ($p > 0.05$). Bacterial growth in PRW was not inhibited, thus indicating no antibacterial effects.

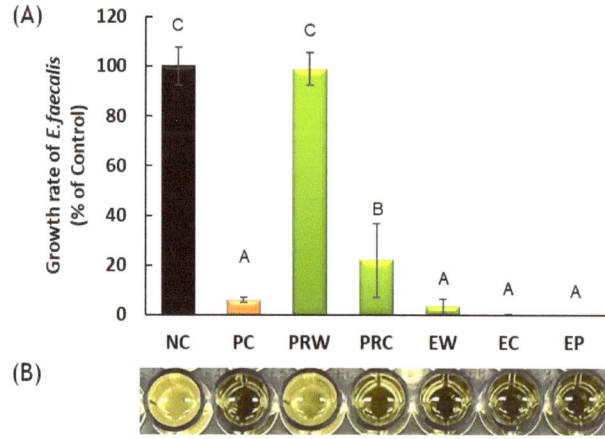

Figure 2. (**A**) Antibacterial activity of dental calcium silicate against *Enterococcus faecalis* by a direct contact test method. Different letters indicate significant differences between groups by one-way ANOVA with Tukey's multiple comparisons ($\alpha = 0.05$). (**B**) Photo of a row of 96-well plate after measuring the optical density. NC: *E. faecalis*, PC: Chlorhexidine. PRW: ProRoot with water, PRC: ProRoot with Chlorhexidine, EW: Endocem with water, EC: Endocem with Chlorhexidine, EP: Endocem Premixed.

3.2. Sustained Antibacterial Effectiveness

Sustained antibacterial activity was evaluated only in the groups that exhibited antibacterial activity (Figure 3). The sustained antibacterial effectiveness of the EP and PRC groups was higher than that of the other experimental groups for up to 30 d ($p < 0.05$). More specifically, on day 20 the EP group showed an *E. faecalis* growth rate of less than 38% of the negative control group and more antibacterial effects than the PRC group ($p < 0.05$). Similarly, the growth rate of the EP group on day 30 was 52%, whereas that of the PRC group was 70%, although there were no statistically significant differences between the two groups ($p > 0.05$).

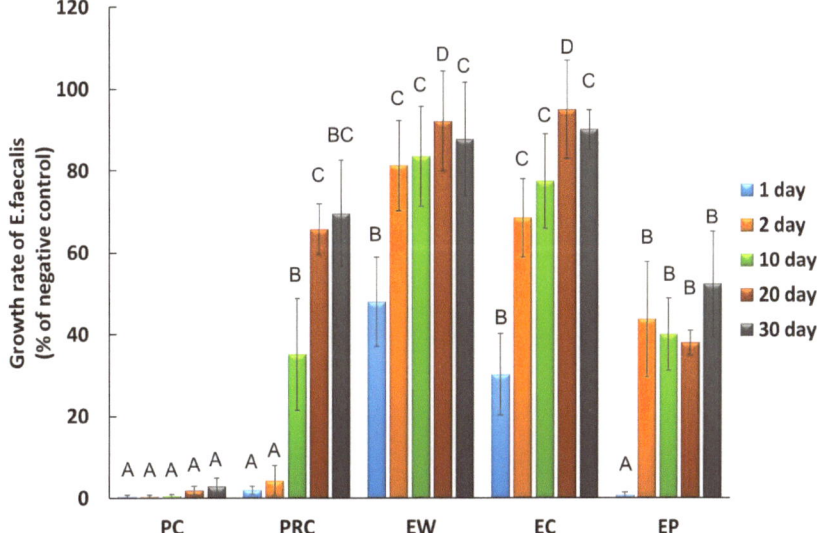

Figure 3. Sustained antibacterial effectiveness of calcium silicate cement against *Entertococcus faecalis*. Values are presented as the percentage of the negative control group that is composed of the bacterial only. Different letters indicate significant differences among the groups on the same day (among the same-colored bars) by one-way ANOVA and Welch with Games-Howell multiple comparisons ($\alpha = 0.05$). Negative control: *E. faecalis* without calcium silicate cement. PC: Chlorhexidine, PRC: ProRoot with Chlorhexidine, EW: Endocem with water, EC: Endocem with Chlorhexidine, EP: Endocem Premixed.

3.3. Cytotoxicity on Human Periodontal Ligament Stem Cell (hPDLSC)

Cytotoxicity was measured after 1, 2, and 3 days of incubation (Figure 4). There were no significant differences between the groups and the NC group ($p > 0.05$) after a day of incubation of the samples. On days 2 and 3, the PRC group exhibited the lowest cell viability among the groups ($p < 0.05$). The cell viability of the PRC group on day 3 was 66%, whereas that of the other groups were all over 70%.

3.4. Scanning Electron Microscopic Analysis

The scanning electron microscopic images of each specimen revealed various surfaces (Figure 5A). The PRC and EC particles were finer and exhibited a more uneven surface texture than those of PRW, EW, and EP. The number of *E. faecalis* attached to each specimen was different and fewer bacteria were observed in the specimens, as in EP (Figure 5B,C). This is consistent with the results of the antibacterial activity and sustained antibacterial effectiveness.

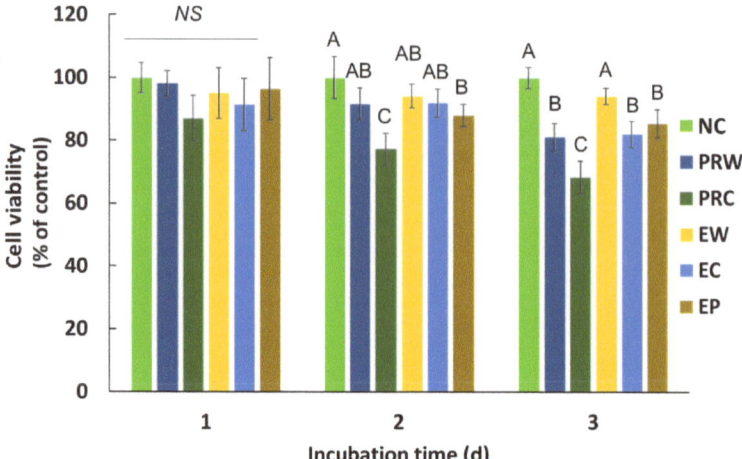

Figure 4. Cell viability on human periodontal ligament stem cells. Different letters indicate significant differences between groups on the same incubation day and NS means no significance by one-way ANOVA with Tukey's multiple comparisons ($\alpha = 0.05$). NC: cell only, PC: Chlorhexidine. PRW: ProRoot with water, PRC: ProRoot with Chlorhexidine, EW: Endocem with water, EC: Endocem with Chlorhexidine, EP: Endocem Premixed.

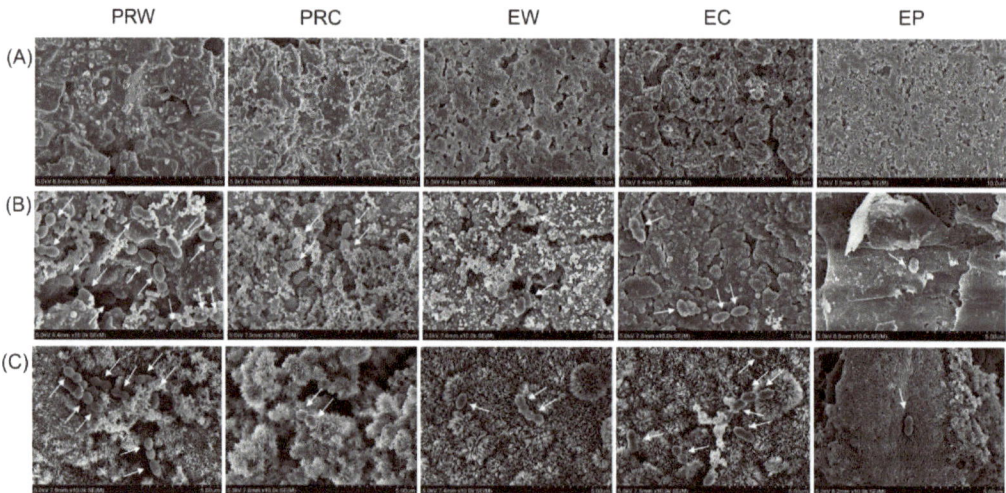

Figure 5. SEM images of (**A**) before the test, (**B**) after the antibacterial test, and (**C**) after the sustained antibacterial effectiveness test on day 10. PRW: ProRoot mixed with water, PRC: ProRoot with Chlorhexidine, EW: Endocem with water, EC: Endocem with Chlorhexidine, EP: Endocem Premixed. Arrows indicate *Enterococcus faecalis* adhesion. (**A**) 5000× magnification, (**B,C**) 10,000× magnification.

3.5. Radiopacity

Among all the specimens tested, the EP group showed the highest radiopacity ($p < 0.05$) (Figure 6). The radiopacities of ProRoot MTA and Endocem MTA were not affected by either of the mixing liquids, i.e., either distilled water or CHX ($p > 0.05$).

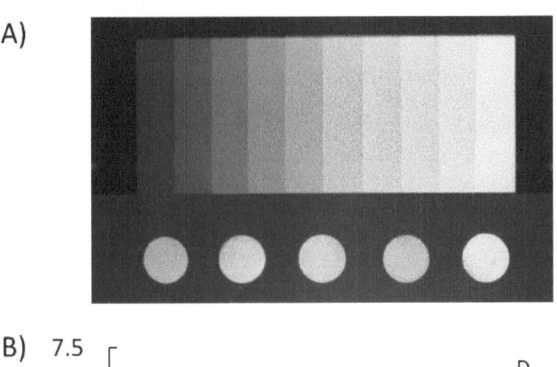
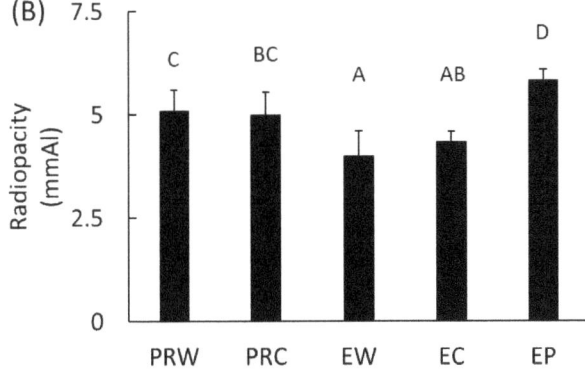

Figure 6. Radiopacity of the calcium silicate cements. (**A**) Representative photo of aluminum step wedge and specimens and (**B**) radiopacity of each group converted to the thickness of aluminum step wedge. Different letters indicate significant differences between groups by one-way ANOVA with Tukey's multiple comparisons ($\alpha = 0.05$). PRW: ProRoot with water, PRC: ProRoot with Chlorhexidine, EW: Endocem with water, EC: Endocem with Chlorhexidine, EP: Endocem Premixed.

4. Discussion

Retrograde filling, also known as apicoectomy or root-end filling, is an endodontic treatment procedure that involves the removal of the apex of the tooth root and its replacement with biocompatible material [23]. When root canal treatment fails, the anatomical complex of the root-end region is removed and retrograde filling is performed to seal the apex and promote healing. An important characteristic of the materials used for retrograde filling is their ability to seal the root-end region and their biocompatibility [18]. Radiopacity of the apical region is also essential as it would allow an evaluation of the status of the applied material through radiography [24]. Ideally, the material should possess antibacterial activity and sustained antibacterial effects against bacteria such as *E. faecalis*, which often leads to the failure of root canal treatment [25]. Therefore, this study aimed to assess the antibacterial activity and sustainability of the antibacterial effect as well as the cell cytotoxicity and radiopacity of injectable premixed CSC and conventional CSCs when mixed with distilled water or CHX.

The antibacterial activities of the EW, EC, and EP groups were superior in comparison to those of the PRC. No antibacterial effects were observed in the PRW group. This may be attributable to differences in the characteristics of the constituents of CSCs [26,27]. CHX is an antiseptic solution used as a mouthwash or for topical treatment to prevent or treat periodontal disease and dental caries. Stowe et al. [28] showed that the antimicrobial properties of CSC were enhanced when mixed with CHX. Numerous other studies have also observed that the antibacterial activities of CSCs improve when mixed with CHX [12,28,29]. Therefore, the use of CHX with CSC in endodontic procedures can improve the sealing ability of CSC, reduce bacterial colonization, and promote faster healing of surrounding tissues [30]. In CSCs with resin, the addition of CHX showed no changes in tensile strength, cytotoxicity, water sorption, or solubility [31]. In this experiment, while CHX enhanced the antibacterial activity of ProRoot, we could not observe any change in antibacterial capability induced by CHX in Endocem as it already had antibacterial properties.

The cell viability of all the groups surpassed 70% of the negative group, with the exception of the PRC group on day 3 which was composed of the cells only. Despite the strong antibacterial effects of EP, EW, and EC, these groups did not exhibit cytotoxicity. A cell viability below 70% of the negative control is to be regarded as cytotoxic, according to the recommendations in ISO 10993-5 [21]. Other studies have established that CSC mixed with CHX increases apoptosis in gingival fibroblasts [10]. Previous studies on CSC primarily used dental pulp stem cells (DPSCs) because CSCs are often used for pulp capping [32,33]. However, in the context of retrograde filling, both DPSCs and PDLSCs play important roles [34]. Therefore, in the present study, we evaluated the cytotoxicity of PDLSCs. Although dental pulp stem cells and PDLSCs belong to the same mesenchymal stem cell lineage, they show differences in proteomics under osteogenic conditions [35]; they exhibit differences in the orientation of the sheet and stem cell markers as well [36]. CSCs used for root-end filling stimulate growth and proliferation of the PDLSCs, leading to improved healing and better tooth prognosis [37].

The sustained antibacterial effectiveness of EP was demonstrated for up to 30 days. Although few studies have investigated the sustained antibacterial effectiveness of CSCs [38,39], no reports have been published on their sustained antibacterial effectiveness beyond 10 days. Odabaş et al. [38] confirmed that the antibacterial activity against *E. faecalis* of CSC mixed with silver zeolite persisted for 72 h. According to Jafari et al. [39], MTA Filapex, a calcium silicate sealer, demonstrated antibacterial activity against *E. faecalis* that persisted for up to 7 days. Few bacteria attached to the surfaces of the EP and PRC specimens were also confirmed by the SEM images (Figure 5). It is worth noting that the EP group continued to exhibit sustained inhibition of *E. faecalis*: less than 38% on the 20th day and 52% on the 30th day.

Several methods to evaluate the antibacterial effects of CSCs have been reported. In a study by Kim et al. [40] that used a disc diffusion test, only EC showed an antibacterial effect on *E. faecalis* 2 h after setting [41]. In another disc diffusion test by Morita et al. [42], assessing the antibacterial activity after setting the CSC was difficult because of the white spots that were formed on the agar plate during the setting. There was a study that evaluated antibacterial activity by directly culturing with *E. faecalis* bacterial broth, however, this study could not identify any differences between the CSC groups [43]. Because the direct-contact test method is considered the most suitable for the comparison of antibacterial activities, it was adopted in this study [19].

The radiopacity of the EP group was the highest and the addition of CHX did not significantly change the radiopacity of ProRoot and Endocem MTA materials. As for the requirements of retrograde filling materials, they should display good radiopacity, possess low solubility, and exhibit appropriate physical properties, such as a low setting time and high compressive strength [27]. Moreover, the addition of CHX or other antibacterial agents should not deteriorate these properties. It is noteworthy that, until now, there are no significant studies related to the long-term physical and mechanical properties, and therefore, further studies are needed to confirm this hypothesis.

Endocem MTA was developed as a quick-setting alternative to address the shortcomings of conventional CSC [41]. Endocem MTA contains fine particles of pozzolan, a silicate-based material that reacts with calcium hydroxide that is formed during cement hydration [44]. As a slew of new products have been developed recently, various studies on their antibacterial activities, cytotoxicity, and physical and mechanical properties are in progress [18]. However, the physical and mechanical characteristics depending on the type of CSCs were not addressed in this study, which should be conducted in subsequent studies. To gain a deeper understanding of the bactericidal and bacteriostatic capabilities of CSCs, there is a subsequent need to conduct measurements using CFU or to perform LIVE/DEAD staining. This study is also limited to in vitro tests. In vivo studies would be required later for clinical applications since a variety of bacteria exist in a dynamic environment in oral conditions.

The antibacterial activity, sustained antibacterial effectiveness, and radiopacity of the fast-set type Endocem and the premixed type Encocem premixed were significantly different from those of the conventional CSC, ProRoot. Therefore, the null hypothesis was rejected. In this study, injectable EP showed excellent and sustained antibacterial activity without cytotoxicity, and the highest radiopacity.

5. Conclusions

In summary, CHX was effective in endowing antibacterial activity and sustained antibacterial effectiveness in the conventional CSC, ProRoot MTA that did not exhibit any antibacterial activity. The addition of CHX did not change the cell viability or radiopacity properties of ProRoot MTA and Endocem MTA, with the exception of cell viability on day 3 in the PRC group.

The pre-mixed injectable type, Endocem premixed, was the most effective CSC, considering its antibacterial activity, sustained antibacterial effectiveness, cytotoxicity, and radiopacity. Endocem premixed showed antibacterial activity and significant antibacterial effect against *E. faecalis* for 30 days with no cytotoxicity compared to other CSC groups. Therefore, it is suggested as a useful antibacterial material for root-end filling.

Author Contributions: S.-H.M.: Data curation and investigation. S.-J.S.: Writing the original draft, conceptualization, and software. S.O.: Methodology, formal analysis, and visualization. J.-M.B.: Writing, review and editing, funding acquisition, supervision, project administration, and validation. All authors have read and agreed to the published version of the manuscript.

Funding: This work was supported by the Korea Medical Device Development Fund grant funded by the Korea government (the Ministry of Science and ICT, the Ministry of Trade, Industry and Energy, the Ministry of Health & Welfare, Republic of Korea, the Ministry of Food and Drug Safety) (Project Number: RS-2020-KD000045).

Institutional Review Board Statement: Not applicable.

Informed Consent Statement: Not applicable.

Data Availability Statement: Data supporting the findings of this study are available from the corresponding author upon request.

Conflicts of Interest: The authors declare no conflict of interest.

References

1. Camilleri, J. Classification of hydraulic cements used in dentistry. *Front. Dent. Med.* **2020**, *1*, 9. [CrossRef]
2. Torabinejad, M.; Watson, T.F.; Pitt Ford, T.R. Sealing ability of a mineral trioxide aggregate when used as a root end filling material. *J. Endod.* **1993**, *19*, 591–595. [CrossRef] [PubMed]
3. Torabinejad, M.; Pitt Ford, T.R. Root end filling materials: A review. *Endod. Dent. Traumatol.* **1996**, *12*, 161–178. [CrossRef] [PubMed]
4. Attik, G.N.; Villat, C.; Hallay, F.; Pradelle-Plasse, N.; Bonnet, H.; Moreau, K.; Colon, P.; Grosgogeat, B. In vitro biocompatibility of a dentine substitute cement on human MG 63 osteoblasts cells: Biodentine™ versus MTA®. *Int. Endod. J.* **2014**, *47*, 1133–1141. [CrossRef] [PubMed]
5. Gomes-Cornélio, A.L.; Rodrigues, E.M.; Salles, L.P.; Mestieri, L.B.; Faria, G.; Guerreiro-Tanomaru, J.M.; Tanomaru-Filho, M. Bioactivity of MTA Plus, Biodentine and an experimental calcium silicate-based cement on human osteoblast-like cells. *Int. Endod. J.* **2017**, *50*, 39–47. [CrossRef]
6. Alghamdi, F.; Shakir, M. The influence of Enterococcus faecalis as a dental root canal pathogen on endodontic treatment: A systematic review. *Cureus* **2020**, *12*, e7257. [CrossRef]
7. Portenier, I.; Waltimo, T.M.T.; Haapasalo, M. Enterococcus faecalis–the root canal survivor and 'star' in post-treatment disease. *Endod. Top.* **2003**, *6*, 135–159. [CrossRef]

8. Vazquez-Garcia, F.; Tanomaru-Filho, M.; Chávez-Andrade, G.M.; Bosso-Martelo, R.; Basso-Bernardi, M.I.; Guerreiro-Tanomaru, J.M. Effect of silver nanoparticles on physicochemical and antibacterial properties of calcium silicate cements. *Braz. Dent. J.* **2016**, *27*, 508–514. [CrossRef]
9. Parirokh, M.; Torabinejad, M. Mineral trioxide aggregate: A comprehensive literature review—Part I: Chemical, physical, and antibacterial properties. *J. Endod.* **2010**, *36*, 16–27. [CrossRef]
10. Hernandez, E.P.; Botero, T.M.; Mantellini, M.G.; McDonald, N.J.; Nör, J.E. Effect of ProRoot MTA mixed with chlorhexidine on apoptosis and cell cycle of fibroblasts and macrophages in vitro. *Int. Endod. J.* **2005**, *38*, 137–143. [CrossRef]
11. Lindblad, R.M.; Lassila, L.V.J.; Vallittu, P.K.; Tjäderhane, L. The effect of chlorhexidine and dimethyl sulfoxide on long-term sealing ability of two calcium silicate cements in root canal. *Dent. Mater.* **2021**, *37*, 328–335. [CrossRef]
12. Mittag, S.G.; Eissner, C.; Zabel, L.; Wrbas, K.T.; Kielbassa, A.M. The influence of chlorhexidine on the antibacterial effects of MTA. *Quintessence Int.* **2012**, *43*, 901–906. [PubMed]
13. Holt, D.M.; Watts, J.D.; Beeson, T.J.; Kirkpatrick, T.C.; Rutledge, R.E. The anti-microbial effect against Enterococcus faecalis and the compressive strength of two types of mineral trioxide aggregate mixed with sterile water or 2% chlorhexidine liquid. *J. Endod.* **2007**, *33*, 844–847. [CrossRef] [PubMed]
14. Stringhini Junior, E.; Dos Santos, M.G.C.; Oliveira, L.B.; Mercadé, M. MTA and biodentine for primary teeth pulpotomy: A systematic review and meta-analysis of clinical trials. *Clin. Oral Investig.* **2019**, *23*, 1967–1976. [CrossRef] [PubMed]
15. Kim, M.; Yang, W.; Kim, H.; Ko, H. Comparison of the biological properties of ProRoot MTA, OrthoMTA, and Endocem MTA cements. *J. Endod.* **2014**, *40*, 1649–1653. [CrossRef]
16. Silva, E.J.N.L.; Carvalho, N.K.; Guberman, M.R.D.C.L.; Prado, M.; Senna, P.M.; Souza, E.M.; De-Deus, G. Push-out Bond Strength of Fast-setting Mineral Trioxide Aggregate and Pozzolan-based Cements: ENDOCEM MTA and ENDOCEM Zr. *J. Endod.* **2017**, *43*, 801–804. [CrossRef]
17. Park, J.H.; Kim, H.J.; Lee, K.W.; Yu, M.K.; Min, K.S. Push-out bond strength and intratubular biomineralization of a hydraulic root-end filling material premixed with dimethyl sulfoxide as a vehicle. *Restor. Dent. Endod.* **2023**, *48*, e8. [CrossRef]
18. Kim, Y.; Lee, D.; Kye, M.; Ha, Y.J.; Kim, S.Y. Biocompatible properties and mineralization potential of premixed calcium silicate-based cements and fast-set calcium silicate-based cements on human bone marrow-derived mesenchymal stem cells. *Materials* **2022**, *15*, 7595. [CrossRef]
19. Eldeniz, A.U.; Hadimli, H.H.; Ataoglu, H.; Orstavik, D. Antibacterial effect of selected root-end filling materials. *J. Endod.* **2006**, *32*, 345–349. [CrossRef]
20. ISO 10993-12; Biological Evaluation of Medical Devices—Part 12: Sample Preparation and Reference Materials. International Organization for Standardization: Geneva, Switzerland, 2012.
21. ISO 10993-5; Biological Evaluation of Medical Devices—Part 5: Tests for In Vitro Cytotoxicity. International Organization for Standardization: Geneva, Switzerland, 2009.
22. Sen, H.G.; Helvacioglu-Yigit, D.; Yilmaz, A. Radiopacity evaluation of calcium silicate cements. *BMC Oral Health* **2023**, *23*, 491. [CrossRef]
23. Paños-Crespo, A.; Sánchez-Torres, A.; Gay-Escoda, C. Retrograde filling material in periapical surgery: A systematic review. *Med. Oral Patol. Oral Cir. Bucal* **2021**, *26*, e422–e429. [CrossRef] [PubMed]
24. Tanomaru-Filho, M.; da Silva, G.F.; Duarte, M.A.; Gonçalves, M.; Tanomaru, J.M. Radiopacity evaluation of root-end filling materials by digitization of images. *J. Appl. Oral Sci.* **2008**, *16*, 376–379. [CrossRef] [PubMed]
25. Rosen, E.; Elbahary, S.; Haj-Yahya, S.; Jammal, L.; Shemesh, H.; Tsesis, I. The invasion of bacterial biofilms into the dentinal tubules of extracted teeth retrofilled with fluorescently labeled retrograde filling materials. *Appl. Sci.* **2020**, *10*, 6996. [CrossRef]
26. Kang, C.M.; Hwang, J.; Song, J.S.; Lee, J.H.; Choi, H.J.; Shin, Y. Effects of three calcium silicate cements on inflammatory response and mineralization-inducing potentials in a dog pulpotomy model. *Materials* **2018**, *11*, 899. [CrossRef] [PubMed]
27. Jang, Y.; Kim, Y.; Lee, J.; Kim, J.; Lee, J.; Han, M.R.; Kim, J.; Shin, J. Evaluation of setting time, solubility, and compressive strength of four calcium silicate-based cements. *J. Korean Acad. Pediatr. Dent.* **2023**, *50*, 217–228. [CrossRef]
28. Stowe, T.J.; Sedgley, C.M.; Stowe, B.; Fenno, J.C. The effects of chlorhexidine gluconate (0.12%) on the antimicrobial properties of tooth-colored ProRoot mineral trioxide aggregate. *J. Endod.* **2004**, *30*, 429–431. [CrossRef]
29. Kapralos, V.; Rukke, H.V.; Ørstavik, D.; Koutroulis, A.; Camilleri, J.; Sunde, P.T. Antimicrobial and physicochemical characterization of endodontic sealers after exposure to chlorhexidine digluconate. *Dent. Mater.* **2021**, *37*, 249–263. [CrossRef]
30. Shahi, S.; Rahimi, S.; Yavari, H.R.; Shakouie, S.; Nezafati, S.; Abdolrahimi, M. Sealing ability of white and gray mineral trioxide aggregate mixed with distilled water and 0.12% chlorhexidine gluconate when used as root-end filling materials. *J. Endod.* **2007**, *33*, 1429–1432. [CrossRef]
31. Vitti, R.P.; Pacheco, R.R.; Silva, E.J.N.L.; Prati, C.; Gandolfi, M.G.; Piva, E.; Ogliari, F.A.; Zanchi, C.H.; Sinhoreti, M.A.C. Addition of phosphates and chlorhexidine to resin-modified MTA materials. *J. Biomed. Mater. Res. B Appl. Biomater.* **2019**, *107*, 2195–2201. [CrossRef]
32. Tomás-Catalá, C.J.; Collado-González, M.; García-Bernal, D.; Oñate-Sánchez, R.E.; Forner, L.; Llena, C.; Lozano, A.; Castelo-Baz, P.; Moraleda, J.M.; Rodríguez-Lozano, F.J. Comparative analysis of the biological effects of the endodontic bioactive cements MTA-Angelus, MTA Repair HP and NeoMTA Plus on human dental pulp stem cells. *Int. Endod. J.* **2017**, *50*, e63–e72. [CrossRef]

33. Tomás-Catalá, C.J.; Collado-González, M.; García-Bernal, D.; Oñate-Sánchez, R.E.; Forner, L.; Llena, C.; Lozano, A.; Moraleda, J.M.; Rodríguez-Lozano, F.J. Biocompatibility of new pulp-capping materials NeoMTA Plus, MTA Repair HP, and Biodentine on human dental pulp stem cells. *J. Endod.* **2018**, *44*, 126–132. [CrossRef] [PubMed]
34. Friedlander, L.T.; Cullinan, M.P.; Love, R.M. Dental stem cells and their potential role in apexogenesis and apexification. *Int. Endod. J.* **2009**, *42*, 955–962. [CrossRef] [PubMed]
35. Kotova, A.V.; Lobov, A.A.; Dombrovskaya, J.A.; Sannikova, V.Y.; Ryumina, N.A.; Klausen, P.; Shavarda, A.L.; Malashicheva, A.B.; Enukashvily, N.I. Comparative analysis of dental pulp and periodontal stem cells: Differences in morphology, functionality, osteogenic differentiation and proteome. *Biomedicines* **2021**, *9*, 1606. [CrossRef]
36. Hu, L.; Zhao, B.; Gao, Z.; Xu, J.; Fan, Z.; Zhang, C.; Wang, J.; Wang, S. Regeneration characteristics of different dental derived stem cell sheets. *J. Oral Rehabil.* **2020**, *47*, 66–72. [CrossRef]
37. Rebolledo, S.; Alcántara-Dufeu, R.; Luengo Machuca, L.; Ferrada, L.; Sánchez-Sanhueza, G.A. Real-time evaluation of the biocompatibility of calcium silicate-based endodontic cements: An in vitro study. *Clin. Exp. Dent. Res.* **2023**, *9*, 322–331. [CrossRef] [PubMed]
38. Odabaş, M.E.; Çinar, C.; Akça, G.; Araz, I.; Ulusu, T.; Yücel, H. Short-term antimicrobial properties of mineral trioxide aggregate with incorporated silver-zeolite. *Dent. Traumatol.* **2011**, *27*, 189–194. [CrossRef]
39. Jafari, F.; Samadi Kafil, H.S.; Jafari, S.; Aghazadeh, M.; Momeni, T. Antibacterial activity of MTA fillapex and AH 26 root canal sealers at different time intervals. *Iran Endod. J.* **2016**, *11*, 192–197. [CrossRef]
40. Kim, R.J.; Kim, M.O.; Lee, K.S.; Lee, D.Y.; Shin, J.H. An in vitro evaluation of the antibacterial properties of three mineral trioxide aggregate (MTA) against five oral bacteria. *Arch. Oral Biol.* **2015**, *60*, 1497–1502. [CrossRef]
41. Morita, M.; Kitagawa, H.; Nakayama, K.; Kitagawa, R.; Yamaguchi, S.; Imazato, S. Antibacterial activities and mineral induction abilities of proprietary MTA cements. *Dent. Mater. J.* **2021**, *40*, 297–303. [CrossRef]
42. Choi, Y.; Park, S.J.; Lee, S.H.; Hwang, Y.C.; Yu, M.K.; Min, K.S. Biological effects and washout resistance of a newly developed fast-setting pozzolan cement. *J. Endod.* **2013**, *39*, 467–472. [CrossRef]
43. Ashi, T.; Mancino, D.; Hardan, L.; Bourgi, R.; Zghal, J.; Macaluso, V.; Al-Ashkar, S.; Alkhouri, S.; Haikel, Y.; Kharouf, N. Physicochemical and Antibacterial Properties of Bioactive Retrograde Filling Materials. *Bioengineering* **2022**, *9*, 624. [CrossRef] [PubMed]
44. Han, L.; Kodama, S.; Okiji, T. Evaluation of calcium-releasing and apatite-forming abilities of fast-setting calcium silicate-based endodontic materials. *Int. Endod. J.* **2015**, *48*, 124–130. [CrossRef] [PubMed]

Disclaimer/Publisher's Note: The statements, opinions and data contained in all publications are solely those of the individual author(s) and contributor(s) and not of MDPI and/or the editor(s). MDPI and/or the editor(s) disclaim responsibility for any injury to people or property resulting from any ideas, methods, instructions or products referred to in the content.

Article

Rheological Properties and Setting Kinetics of Bioceramic Hydraulic Cements: ProRoot MTA versus RS+

Arne Peter Jevnikar [1], Tine Malgaj [2], Kristian Radan [3], Ipeknaz Özden [4,5], Monika Kušter [4,5] and Andraž Kocjan [4,*]

1. Endodent d.o.o., Metelkova ulica 15, 1000 Ljubljana, Slovenia
2. Department of Prosthodontics, Faculty of Medicine, University of Ljubljana, Hrvatski trg 6, 1000 Ljubljana, Slovenia
3. Department of Inorganic Chemistry and Technology, Jožef Stefan Institute, Jamova 39, 1000 Ljubljana, Slovenia
4. Department for Nanostructured Materials, Jožef Stefan Institute, Jamova 39, 1000 Ljubljana, Slovenia
5. Jožef Stefan International Postgraduate School, Jamova 39, 1000 Ljubljana, Slovenia
* Correspondence: a.kocjan@ijs.si; Tel.: +386-14773271

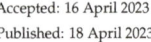

Citation: Jevnikar, A.P.; Malgaj, T.; Radan, K.; Özden, I.; Kušter, M.; Kocjan, A. Rheological Properties and Setting Kinetics of Bioceramic Hydraulic Cements: ProRoot MTA versus RS+. *Materials* **2023**, *16*, 3174. https://doi.org/10.3390/ma16083174

Academic Editors: Luigi Generali, Vittorio Checchi and Eugenio Pedullà

Received: 4 April 2023
Revised: 13 April 2023
Accepted: 16 April 2023
Published: 18 April 2023

Copyright: © 2023 by the authors. Licensee MDPI, Basel, Switzerland. This article is an open access article distributed under the terms and conditions of the Creative Commons Attribution (CC BY) license (https://creativecommons.org/licenses/by/4.0/).

Abstract: Hydraulic calcium silicate-based cements (HCSCs) have become a superior bioceramic alternative to epoxy-based root canal sealers in endodontics. A new generation of purified HCSCs formulations has emerged to address the several drawbacks of original Portland-based mineral trioxide aggregate (MTA). This study was designed to assess the physio-chemical properties of a ProRoot MTA and compare it with newly formulated RS+, a synthetic HCSC, by advanced characterisation techniques that allow for in situ analyses. Visco-elastic behaviour was monitored with rheometry, while phase transformation kinetics were followed by X-ray diffraction (XRD), attenuated total reflectance Fourier transform infrared (ATR-FTIR), and Raman spectroscopies. Scanning electron microscopy with energy-dispersive spectroscopy, SEM-EDS, and laser-diffraction analyses was performed to evaluate the compositional and morphological characteristics of both cements. While the kinetics of surface hydration of both powders, when mixed with water, were comparable, an order of magnitude finer particle size distribution of RS+ coupled with the modified biocompatible formulation proved pivotal in its ability to exert predictable viscous flow during working time, and it was more than two times faster in viscoelastic-to-elastic transition, reflecting improved handling and setting behaviour. Finally, RS+ could be completely transformed into hydration products, i.e., calcium silicate hydrate and calcium hydroxide, within 48 h, while hydration products were not yet detected by XRD in ProRoot MTA and were obviously bound to particulate surface in a thin film. Because of the favourable rheological and faster setting kinetics, synthetic, finer-grained HCSCs, such as RS+, represent a viable option as an alternative to conventional MTA-based HCSCs for endodontic treatments.

Keywords: hydraulic calcium silicate cements (HCSCs); mineral trioxide aggregate (MTA); bioactive materials; handling; rheological properties; setting kinetics

1. Introduction

In the past decade, hydraulic calcium silicate-based cements (HCSCs) have become good alternatives to epoxy-based root canal sealers in endodontics. This group of materials has been shown to possess good bioactivity, which enhances biological healing processes [1]. In addition to sealing ability, hydraulic cements are the material of choice in various clinical indications, including closure of open apices [2], retrograde root canal obturation in apical surgery, direct/indirect pulp capping [3], and repair of internal and external root resorptions [4,5]. Moreover, due to their proven biological properties, hydraulic cements are successfully used in the repair of the furcal perforations, in which it is difficult to control moisture content [6]. In addition to favourable biological and physical properties, clinicians require easy handling characteristics of the material in everyday routine. One of the first bioactive cements used clinically for more than two decades is mineral trioxide aggregate (MTA), developed by Torabinejad et al. [7]. MTA has been shown to induce mineralisation beneath exposed pulp and was initially used for root perforations and as a root end filling material [8,9]. Due to evidence-based clinical success [10–13], the indications for the use of MTA have grown considerably. Until recently, a variety of MTA-based materials was introduced into clinical practice]. However, the ideal endodontic repair material should be dimensionally stable, non-resorbable, radiopaque, biocompatible, and bioactive, and it should have good handling properties [1].

MTA is a Portland cement-based endodontic material made originally by combining grey Portland cement with a radiopaque, brownish-coloured bismuth oxide (Bi_2O_3) powder. The principal compounds of MTA are tricalcium silicate (C3S—Ca_3SiO_5) and dicalcium silicate (C2S—Ca_2SiO_4), but MTA also consists of considerable amounts of other oxides, such as tricalcium aluminate, calcium sulphate, tricalcium oxide, and iron oxide, which are responsible for final physio-chemical properties [14,15]. The original MTA was grey and often caused tooth discolouration. In 2002, "tooth-coloured" white MTA was introduced. Both materials have similar compositions; however, white MTA was reported to have fewer iron oxide traces and a finer particle size [16].

Despite evidence-based clinical success [10–13], several limitations have been reported with MTA, including poor handling characteristics, long setting time [17–19], difficult retrieval from the operation area [20], and post-treatment tooth discoloration [21,22]. Since it is based on Portland cement, an additional concern with respect to clinical employment was in its purity regarding the traces of heavy metals present in MTA-based materials, such as arsenic, bismuth, cadmium, chromium, copper, iron, lead, manganese, nickel, and zinc [23–25]. The sandy nature, resulting in a lack of uniformity of the premixed MTA paste, has caused some clinical limitations in paste application. Since the introduction of MTA, one of its greatest drawbacks has been the long setting time. It has been shown that MTA sets slowly in approximate 3–4 h [16,26,27]. Clinicians require a material that would ideally set before the end of the procedure.

In light of the above-mentioned drawbacks, original MTA-type formulations have been modified, offering new, purified (and/or synthetic) HCSCs systems, including Biodentine, Bioaggregate, Endosequence, and others [1,28,29]. The common material evolution was in the simplification of the compositions, while also increasing their biocompatibility. For example, Bi_2O_3 was replaced with biomedical grade zirconia (3Y-TZP), while the active ingredient C3S (with or without C2S) was coupled with calcium carbonates, silicates, or phosphates to promote setting and/or remineralisation [27,30]. Recently, RS+ bioceramic root canal repair material, a product from Jožef Stefan Institute's spin-off company, has been introduced to the dental community, claiming improved handling and biological properties compared to the original MTA formulations. It is based on synthetic C3S and zirconia as a radioopacifier, with small additions of biocompatible phyllosilicate clay (bentonite) and bioactive amorphous calcium silicate for the enhanced handling, setting, and remineralisation properties of powder-like materials. The bioactive amorphous calcium silicate was shown to be a superior osteoinductive compared to calcium phosphates, such as beta-calcium phosphate and hydroxyapatite, which are commonly added to similar formulations [31,32].

Clinically, the setting time of cement paste is defined by the transition time from a fluid state into a solidified state [33]. Setting time can be affected by mixing method, quantity of water used, packaging force, and moisture of the environment. In addition to the needle penetrations using Gilmore weights, as defined by the ISO 6876 and ISO 9917-1 standards to determine setting time [34], there are several other means to experimentally evaluate setting time, including microhardness and strength measurements [35]. These methods provide some information about the setting characteristics; however, the setting reaction is observed indirectly. Moreover, handling, another very important property of successful clinical application, has rarely been tested and defined.

Handling properties and viscoelastic-to-elastic transition during setting can be precisely measured by rheometry in rotational and oscillatory modes, respectively. In situ X-ray diffraction, Fourier transform infrared (FTIR), and Raman spectroscopies can continuously follow the setting mechanism of hydraulic cements, which can in turn estimate the amounts of the constituents in samples and the transition kinetics. The purpose of this study was to evaluate the physio-chemical properties, in terms of handling and setting behaviours, of a benchmark MTA formulation ProRoot MTA hydraulic cement (Dentsply Tulsa Dental, Tulsa, OK, USA), and to compare them with those of newly formulated RS+ (Genuine Technologies d.o.o.; Ljubljana, Slovenia), by advanced characterisation techniques that allow for in situ observations of the phase and visco-elastic transformation kinetics, such as X-ray diffraction (XRD), rheometry, FTIR, and Raman spectroscopies. In addition, scanning electron microscopy with energy-dispersive spectroscopy (SEM-EDS) and laser-diffraction analyses were performed to evaluate the compositional and morphological characteristics of both cements affecting the physio-chemical properties. The null hypothesis was that there are no differences in compositional or morphological characteristics, in hydration and phase transformation kinetics, and in rheological behaviour between ProRoot MTA and a novel RS+.

2. Materials and Methods

2.1. Materials and Sample Preparation

The two hydraulic cements analysed in this research (ProRoot MTA and RS+) were prepared according to the manufacturers' instructions. When mixing the ProRoot MTA, a new pouch of material was opened and dispensed on a mixing pad, followed by a ProRoot MTA liquid ampule being opened and its contents being squeezed out next to the material. The powder and the liquid were gradually incorporated using a mixing spatula until a thick, creamy consistency was formed. In the case of RS+, 0.3 g of powder was mixed with six drops of deionised water. The powder was first weighed on an electronic scale and then transferred to a mixing plate. Drops of distilled water were placed, and next, the powder and the liquid were incorporated using a mixing spatula until the desired consistency was obtained. As-prepared paste was then transferred to the given measuring cell (XRD, FTIR). Measurements were performed at room temperature. In the case of hydration experiments, the paste consistency of cement immediately after mixing were allowed to set over 48 h at 37.5 °C and saturated under humidity in an incubator. Afterwards, the hydrated cement was crushed in a mortar and analysed with XRD. The experimental flowchart is presented in Figure 1.

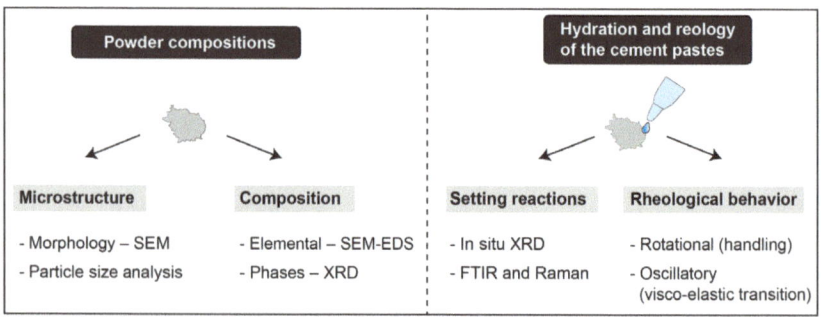

Figure 1. Experimental flowchart.

2.2. Scanning Electron Microscopy Coupled with Energy-Dispersive Spectroscopy (SEM-EDS)

As-received powders were analysed by SEM-EDS to evaluate their particulate morphology and elemental composition. SEM analyses were performed using an FEI Helios NanoLab focused ion beam (FIB)-SEM (Helios Nanolab 650, FEI, Hillsboro, OR, USA). Measurements of chemical composition were performed with an 50-mm^2 X-max silicon drift detector (SDD) (Oxford Instruments, Abingdon, UK) attached to the microscope. Prior to the analysis, the powders were mounted on a carbon tape and sputtered with carbon. The SEM imaging was performed at an accelerating voltage and a beam current of 5 kV and 0.1 nA, respectively. EDS mapping was performed at a 15-kV accelerating voltage and 10-µs dwell time to obtain qualitative visualisation of the distribution of elements greater than the detection limit within the particulate samples.

2.3. Particle Size Distribution

The particle size distributions (PSDs) of both powders and their mixtures were determined using the laser diffraction method (Horiba LA-920, Kyoto, Japan). To avoid hydration reactions, 2-propanol was used as a solvent. Prior to the measurements, the non-aqueous powder dispersions were rigorously mixed and ultrasonicated for 5 min in the measuring cell. Volume and cumulative volume particle distributions were plotted.

2.4. XRD Phase Characterisation

The X-ray diffraction (XRD) data from (un)hydrated powders were collected with a Malvern Panalytical Empyrean X-ray diffractometer (Empyrean multipurpose X-ray diffractometer, Almelo, The Netherlands) using a monochromated X-ray beam produced by a Cu-target tube ($\lambda K\alpha 1$ = 0.15406 nm and $\lambda K\alpha 2$ = 0.154439 nm). The measurements were obtained in Bragg-Brentano geometry using a range of $10° < 2\Theta < 80°$, a step size of $0.0131°$, a divergence slit of 0.04 rad, and a counting time of 1 s per step. In addition, the in situ hydration measurements were performed on Bragg-Brentano geometre in a range of $41° < 2\Theta < 42°$, applying a step size of $0.0131°$ and using a divergence slit of 0.04 rad and a counting time of 1 s per step. On the diffracted part, the large β-Ni filter was used to reduce the intensity of the $K\alpha 2$ line. After mixing of the powder with the deionised water to produce a suitable paste, the latter was spread onto the in situ holder and measured at different times. In situ experiments collected data at different times at 2 min, 5 min, 10 min, 15 min, 20 min, 25 min, 30 min, 40 min, 1 h, and 2 h. The XRD data were analysed using HighScore Plus XRD Analysis Software database PDF-4+.

2.5. Rheology

Rheological characterisation of the cement pastes was performed in rotational, as well as oscillatory tests, using a Physica MCR 301 rheometer (Anton Paar GmbH, Graz, Austria). In rotational mode, the change in the viscosity of cement pastes was measured for both samples at a constant temperature of 25 °C and a constant shear rate of 10 s^{-1} for 20 min with a plate-on-plate system with a 15-mm upper plate. The setting kinetics were monitored by oscillatory measurements following the real-time build-up of the storage modulus G' (Pa) under a sinusoidal strain of 0.0005% with an oscillation frequency of 1 radian per second. During the measurements, the plate-on-plate system with a 15-mm upper plate was used, and a gap of 0.051 mm was set between the two plates. A constant temperature of 25 °C was maintained for the duration of the measurements.

2.6. Infrared Spectroscopy

The measurements were performed using a PerkinElmer Spectrum 100 Fourier transform infrared (FT-IR) spectrometer (PerkinElmer, Waltham, MA, USA) equipped with a universal attenuated total reflectance (UATR) module. After an air background calibration, the freshly mixed cement pastes were placed directly on the optic window of the diamond ATR top plate fitted with a metal sample holder (powder cup). Each paste was loaded in the sample compartment, where it was closed and compressed by the pressure applicator tip with enough pressure to ensure good and uniform contact between the diamond and the sample. The spectra were collected at different times (soon after mixing and at 1.5 min, 4 min, 6 min, 8 min, 10 min, 12 min, 14 min, 20 min, 25 min, 30 min, 40 min, 45 min, 50 min, 1 h, 1.5 h, 2 h, 3 h, 4 h, 5 h, 7 h, 10 h, 72 h, and 76 h after mixing) in the 300–4000 cm^{-1} range, with a spectral resolution of 4 cm^{-1} as average spectra out of 16 scans. The spectrometer was running Spectrum 10 software (PerkinElmer, Waltham, MA, USA).

2.7. Raman Spectroscopy

Spectra were collected at room temperature using a Horiba Jobin-Yvon LabRam HR confocal Raman system, coupled with an Olympus BXFM-ILHS microscope. Samples were excited by the 633-nm emission line of a He−Ne laser with a power output of 7 mW on the sample through a 10× microscope objective. Thirty scans per sample were averaged over the spectral range of 100–4000 cm^{-1} with a spectral resolution of 2 cm^{-1} and exposure time of 1 s. The cement pastes were prepared directly on the microscope glass slides and measured at the same time intervals as for the FT-IR spectroscopic measurements.

3. Results

3.1. Morphological and Particle Size Analyses

As-received powders were inspected with SEM to observe their respective particulate morphologies (Figure 2). As seen on a lower-magnification SEM micrograph (Figure 2a) ProRoot MTA powder is composed of irregularly shaped particles with a broad size distribution ranging from one to several tens of micrometres in size. A higher-magnification SEM micrograph (Figure 2a) revealed that agglomerated clusters of finer particles of irregular shapes were attached to the larger particles. Particles resembled sharp edges and vertices. RS+ powder, on the other hand, was substantially finer as seen from the lower-magnification SEM micrograph (Figure 2b). It was only possible to assess the particulate morphology from the higher magnification micrograph, on which it can be seen that larger, micron-sized cuboid particles were covered with finer, 100 nm-sized spherical particles. The larger particles correspond to C3S, while the smaller match the size of zirconia (3Y-TZP) and amorphous bioactive calcium silicate.

Figure 2. SEM micrographs showing the powder morphology, agglomeration state, and particle size: (**a**) ProRoot MTA and (**b**) RS+.

Laser diffraction was used to analyse the particle size distributions (PSDs) of both cements (Figure 3). ProRoot MTA had a broader monomodal PSD with mean particle size (d_{mean}) of approx. 15 μm, in accordance with SEM (Figure 2a). RS+ had a narrower monomodal PSD distribution with an order of magnitude-finer d_{mean} value of 1 μm. The broadness of the PSD can be depicted by d_{10}, d_{50}, and d_{90} from cumulative volume distributions. ProRoot MTA exhibited d_{10}, d_{50}, and d_{90} of 3.2, 11.0, and 28.2 μm, respectively. RS+ resembled d_{10}, d_{50}, and d_{90} of 0.5, 0.9, and 1.9 μm, respectively. The broadness of the PSD is reflected in the magnitude between d_{10} and d_{90}. In the case of ProRoot MTA, d_{90} was more than seven times higher than d_{10}, while in the case of RS+, this increase was four fold.

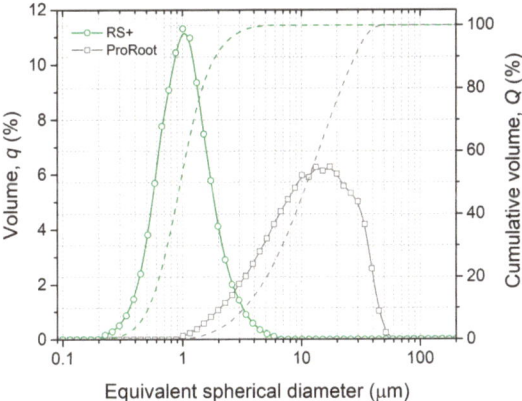

Figure 3. Particle size distributions of as-received hydraulic cements.

3.2. Elemental Composition

SEM-EDS mapping analysis of both cements was performed to observe the elemental distribution (Figure 4). The major elements in both cements are, as expected, calcium and silicon due to C3S being the major phase and, in the case of ProRoot MTA, owing to the C2S presence as well. In ProRoot MTA, large, micrometre-sized regions of bismuth are visible, indicating the size of Bi_2O_3 particulate inclusions as radiopacifying agents. In the case of RS+, zirconia is used and is distributed homogeneously throughout the sample owing to the much smaller particles (Figure 2b). ProRoot MTA contained several other elements at greater the detection limit, i.e., sulphur, aluminium, and iron, corresponding to calcium sulphate, tricalcium aluminate, and iron oxide, respectively. It is interesting to note the variation in the concentration of calcium in RS+.

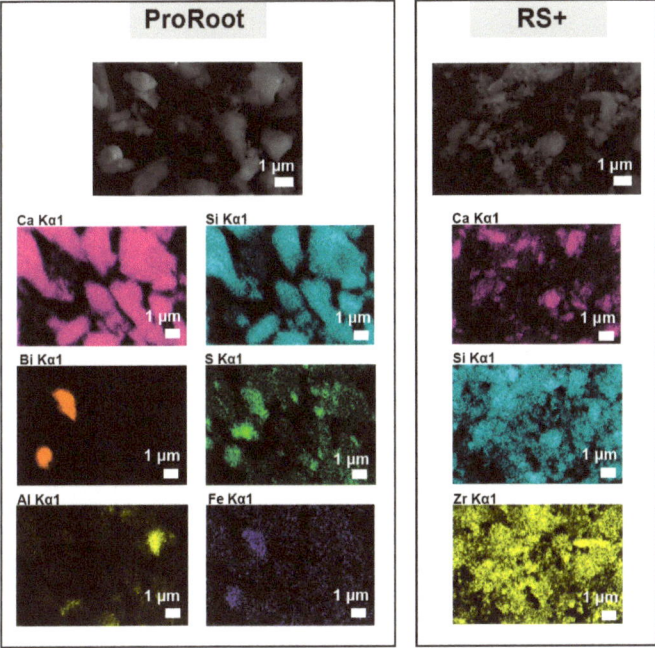

Figure 4. SEM-EDS mapping of the as-received powders showing elemental distribution.

3.3. XRD Phase Composition

Unhydrated and 48-h hydrated cement powders were analysed by XRD. In Figure 5a, the diffractogram of as-received ProRoot MTA indicates the presence of several main phases. These phases are bismuth oxide, tricalcium silicate, dicalcium silicate, and calcium sulphate, in accordance with the EDS-SEM analysis (Figure 4). Calcium aluminate was not detected, indicating its presence in traces, i.e., less than a few wt.%. Surprisingly, after hydration of ProRoot MTA, no visible changes in phase composition were detected, except for the change in the relative intensities of the peaks. In the case of RS+ powder, the XRD diffractograms before and after hydration are presented in Figure 6. As-received powder (Figure 6a) was only composed of zirconia (tetragonal and monoclinic polymorphs) and C3S. Hydrated RS+ powder yielded more significant changes compared to the case of ProRoot MTA. The C3S was completely replaced with the new phases, i.e., portlandite ($Ca(OH)_2$) and calcium silicate hydrate (CSH) as main hydration products, indicating the completed hydration reaction in 48 h. In addition, traces of calcium carbonate were also detected. Compared to ProRoot MTA, the diffractogram of the as-received RS+ powder indicated the finer crystallinity and the presence of amorphous phases (due to the addition of bioactive calcium silicate) due to broader diffraction peaks and the higher background, respectively.

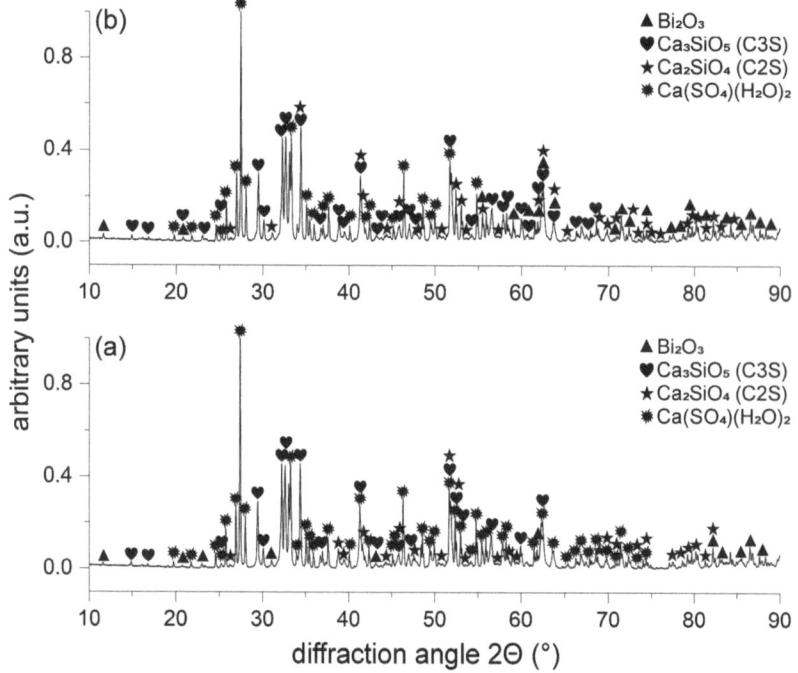

Figure 5. XRD diffractograms of the ProRoot MTA: (**a**) as received; and (**b**) after 48-h hydration.

The higher reactivity of the RS+ powder allowed for in situ XRD measurements of the hydration kinetics in the 41–41.5° 2θ range (Figure 6b). The hydration reaction was monitored by observing (012) the peak of C3S positioned at 41.148° 2θ (PDF 00-031-0301). While there were no significant changes in the intensity of the (peak) after 5–40 min of hydration in the XRD setup, the diffraction peak was gone after 120 min, indicating the depletion of the phase with the hydration reaction.

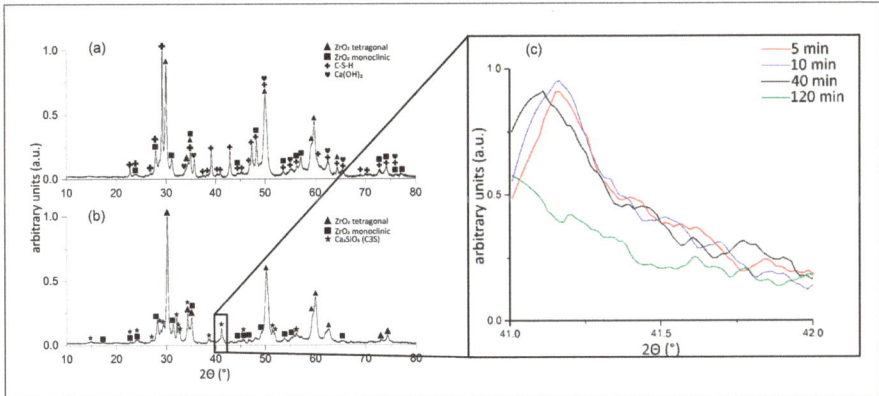

Figure 6. XRD diffractograms of the RS+ CSH cement: (**a**) as received; and (**b**) after hydration. (**c**) In situ XRD monitoring of the hydration kinetics in the 41–41.5 2θ range.

3.4. FTIR and Raman Spectroscopies

FTIR was employed to monitor the surface changes in the cements during the hydration process (Figures 7 and 8). Due to the strong overlap of the broad band of H_2O vibration and absence of clear time-dependent changes less than 1000 cm^{-1}, the 1300–1800 cm^{-1} and 2600–3800 cm^{-1} regions were selected and expanded for the study of the early stages of cement hydration. Both hydraulic cements, ProRoot MTA (Figure 7), and RS+ (Figure 8), exhibited similar spectral evolution during the hydration process, as evidenced by the decrease in the intensities of the characteristic broad water O–H stretching absorption in the range of 2800–3700 cm^{-1} and of the sharper H–O–H bending mode at approximately 1640 cm^{-1} [36]. The change in the aforementioned bands was minimal in the first 14 min following the mixing with mixing solution or deionised water, but a rapid decrease in the intensities can be observed at up to 40 min, after which the associated spectra are virtually identical.

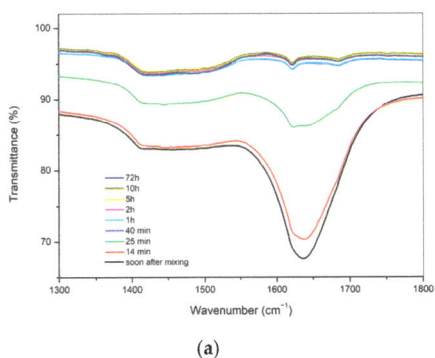
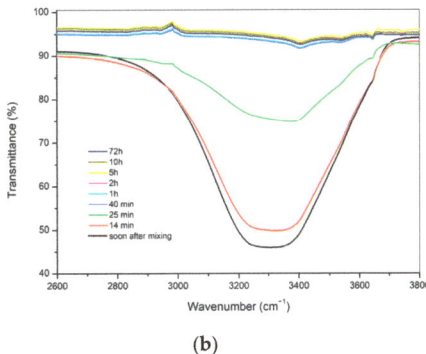

(**a**) (**b**)

Figure 7. ATR-FTIR spectra of the ProRoot MTA paste at selected time intervals following mixing shown in the regions (**a**) 1300–1800 cm^{-1}; (**b**) 2600–3800 cm^{-1}.

In both analysed materials, ProRoot MTA and RS+, the decrease in intensity of the water bands is paralleled by the appearance of a band centred around 3640 cm^{-1}, which can be ascribed to the O–H stretching mode of $Ca(OH)_2$ (portlandite). At the onset of the hydration reaction, this band appears as a weak shoulder protruding from the broad H_2O region, and over time, its intensity increases while becoming better resolved.

The time-dependent Raman spectra of ProRoot MTA and RS+ aqueous cement pastes in the 50–4000 cm^{-1} range are shown in Figure 9. ProRoot spectra exhibit sharp bands with high intensities in the 50–1000 cm^{-1} region, where the majority of all vibration modes are present. An interesting feature of ProRoot MTA is also the absence of O–H stretching vibrations of water molecules in the 3000–3800 cm^{-1} region, while a broad corresponding band can be clearly observed in RS+ soon after mixing. Due to the high spatial resolution of the method, the Raman measurements were repeated on a second RS+ paste to assess the average response of the sample, and the general spectral features were found to be highly comparable at all time intervals. The most prominent time-dependent change in the RS+ sample is represented by the band at 1083 cm^{-1}, which arises soon after mixing and becomes the dominant signal approximately after 8 min following the mixing. This band can be assigned to the formation of different phases of $CaCO_3$, namely to the symmetric (v_1) stretching of the C–O bonds, and it indicates a rapid uptake of CO_2 upon exposure of the pasters to air [37]. Traces of $CaCO_3$ were also detected with XRD. No evidence of the carbonation can be observed in ProRoot MTA, where the appearance of the band at 1008 cm^{-1} might be attributed to the symmetrical stretching of Q^2 silicate units (chains), $v_1(SiO_4)$ [38], which appears soon after mixing but does not grow in intensity over time. Both the FTIR and Raman spectra of ProRoot MTA obtained in this study are similar to those reported by other research groups [39–41], with differences in the degrees of cement hydration and atmospheric carbonation due to different sample preparations and measuring setups used.

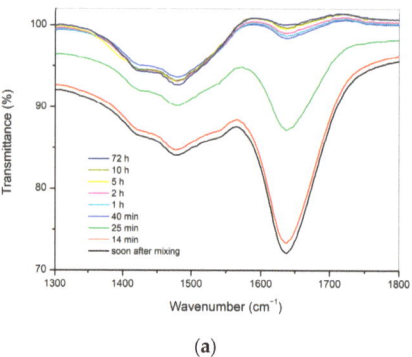

(a)

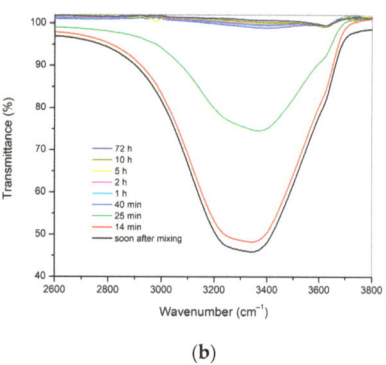

(b)

Figure 8. ATR-FTIR spectra of the RS+ paste at selected time intervals following mixing shown in the regions of: (**a**) 1300–1800 cm^{-1}; and (**b**) 2600–3800 cm^{-1}.

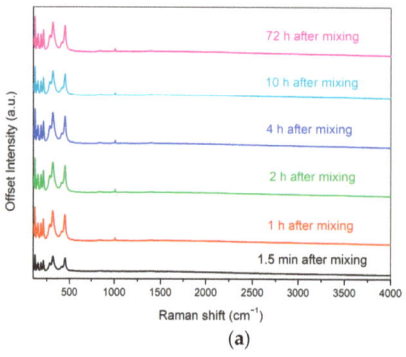

(a)

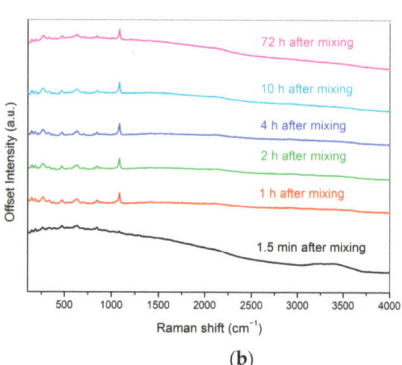

(b)

Figure 9. Raman spectra of the smeared ProRoot MTA (**a**) and RS+ (**b**) samples at selected time intervals following mixing. Individual spectra have been vertically offset for clarity.

3.5. Rheological Properties

Rheological property measurements of the cement pastes immediately after hydration were performed in rotational, as well as in oscillatory, rheometry to assess viscosities and viscoelastic-to-elastic transition, respectively. In rotational mode (Figure 10a), the change in the viscosity (at a constant shear rate) with time provides a direct indication of the setting hydration reactions. In the case of RS+ aqueous pastes, the initial viscosity decreased from 911.8 to 473.3 Pa·s in the first 2–3 min During this "working" period, it exhibited a shear-thinning behaviour typical for thick ceramic suspensions [42]. Afterwards, the viscosity started to steadily increase, becoming shear thickening and indicating the hydration setting reactions. After 17.5 min, soon after reaching the highest viscosity value of 24,050 Pa·s, a sudden drop occurred that corresponded to the sample solidification and (partial) detachment from the measuring cell. On the other hand, it was not possible to obtain a legitimate flow curve behaviour for ProRoot MTA paste. The repetitive increase and decrease in the viscosity were due to the inhomogeneous, sandy-like paste state, which was unable to exhibit a viscous fluid-like flow. The reason for this outcome might be a relatively large PSD (Figure 3) and/or partial solidification of the material.

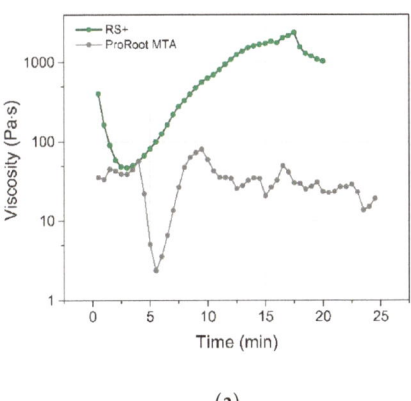
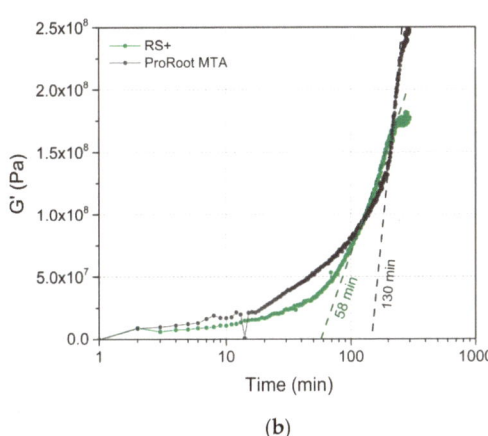

(a) (b)

Figure 10. (a) Viscosity of RS+ and ProRoot MTA plotted as a function of time at the fixed shear rate; (b) storage modulus of RS+ and ProRoot MTA plotted as a function of time.

The oscillatory mode was used to observe the viscoelastic properties of both pastes during setting and to observe viscoelastic-elastic transition (Figure 10b). The material is initially viscoelastic, in which the interlocking of cement begins to occur on account of the precipitation of crystalline products. When this process occurs, the storage shear elastic modulus (G′) starts to rise. At this stage, the cement mix is becoming progressively thicker and is less able to flow. In both cases, an exponential increase in storage shear elastic modulus (G′) was observed. In the case of ProRoot MTA the increase in G′ was not as consistent as in RS+ and was composed from at least two different regions, indicating more complex solidification. The final values of G′ were 1.77×10^8 and 2.48×10^8 Pa for ProRoot MTA and RS+, respectively, indicating higher stiffness of ProRoot MTA. Extrapolating the slope of the maximum increase in G′ provided a rough estimation of the time at which the transition from viscoelastic to elastic material occurred and could be considered as the time of solidification (setting), i.e., as a result of more comprehensive particle interlocking [33]. This transition occurred more than two times faster in the case of RS+ compared to ProRoot MTA (60 versus 120 min). The difference in attained higher stiffness for the ProRoot MTA could be ascribed to the material being still largely unreacted after 48 h of hydration and also consisting of a much larger PSD, providing a stiffer matrix connected with the thin film of hydrated products.

4. Discussion

In this study, differences in compositional and morphological characteristics, in hydration and phase transformation kinetics and rheological behaviour between ProRoot MTA and a novel RS+ were detected. Therefore, the null hypothesis was rejected. RS+ is considered a next generation synthetic HCSC with a refined physio-chemical formulation for improved handling, setting, and biological properties, while ProRoot MTA is a benchmark MTA material extensively studied and used clinically.

The morphologies and particle size distributions (PSDs) of both HCSCs were different. RS+ had a narrower mono-modal PSD with an order of magnitude-finer d_{mean} value compared to ProRoot MTA, i.e., 1 µm versus 15 µm (Figure 3). ProRoot MTA particles, in addition to being larger, also resemble sharp edges and vertices (Figure 2), which could be related to the milling of clinker in Portland cement production [43]. RS+ powder, on the other hand, is substantially finer since it is produced synthetically, with micron-sized cuboid particles covered with finer, spherical-shaped particles (Figure 2b). Further distinctions between the HCSC powders were revealed with SEM-EDA mapping and XRD. The former confirmed a more complex composition of ProRoot MTA (Figure 4). Bismuth was detected, and in addition to calcium and silicon, sulphur, aluminium, and iron were detected as well. Bismuth oxide, used as the main radioopacifier, has been reported to be cytotoxic, negatively affecting cell proliferation [23]; thus, the newer HCSCs primarily use other biocompatible radiopaque materials, for example, biomedical grade zirconia [1], as is the case for RS+ as well. The XRD analysis confirmed the much more complex and diverse phase composition of ProRoot MTA (Figure 5) compared to RS+ (Figure 6a). In the case of the latter, homogeneously distributed calcium, silicon, and zirconium were detected. The variation in the concentration of calcium could be attributed to the distribution or agglomeration of other calcium-containing compounds present in the RS+ formulation, such as amorphous calcium silicate and/or bentonite.

The kinetics of surface hydration for both powders when mixed with water were analogous and comparable, as evidenced by the evolution of the time-dependent ATR FT-IR spectra (Figures 6 and 7). Hydration reaction kinetics occur on the surface of the particles. The difference in PSDs between both HCSCs is reflected in their Raman spectra, where the RS+ paste is characterised by much broader, weaker, and less distinct bands in comparison with ProRoot MTA. This difference is typically observed in amorphous materials with short-range ordering. Moreover, strong carbonate bands arising from surface carbonation observed in the RS+ spectra but were absent in the case of ProRoot MTA, which might also be related to its fine-grained matrix, resulting in greater surface area and thus accelerating the atmospheric CO_2 mineralisation with portlandite formed during the setting reaction.

The differences in chemical composition, an order of magnitude-finer PSD of RS+ coupled with the more uniform particulate morphology and addition of plasticizer (bentonite), proved pivotal in dictating the setting dynamics, namely in the ability to exert predictable viscous flow during working time (Figure 10a) and more than two times faster viscoelastic-elastic transition (Figure 10b), reflecting improved handling and setting behaviour. Bentonite is known for its ability to adsorb significant amounts of ions affecting the bentonite-containing slurries [44], high swelling capacity, and the formation of a gel-like structure with yield characteristics and viscoelastic properties at relatively low phyllosilicate clay concentrations [45,46]. Longer setting times for ProRoot MTA and the inhomogeneous sandy-like paste state are in agreement with the previous studies [16,26,27], in which even longer times of up to 300 min were reported to achieve a higher shear modulus [27].

RS+ could be completely transformed into hydration products, i.e., calcium silicate hydrate and calcium hydroxide, within 48 h, while hydration products were not yet detected by XRD in ProRoot MTA. In the case of the latter, the hydration was obviously bound to particulate surface thin film connecting the matrix of the unreacted particles, indicating that the hydration reactions were ongoing, albeit at a slower rate. The formed surface reaction film was responsible for an immediate and continuous increase in G′

for ProRoot MTA (Figure 10b), indicating ongoing solidification, in accordance with the previous findings [26,27].

Difficult handling is one of the main drawbacks of conventional MTA-based HCSCs. The inability to exert flow and slow hydration bound to the surface might be the reasons for the difficult adaptation of MTA materials in the root canal system or other defects [47]. An improved material's rheological behaviour could represent a clinical advantage. A predictable viscous flow provides easier handling and material adaptation, while higher viscosity decreases the probability of material dislodgement during its application, such as periapical extrusion [47]. On the other hand, faster setting kinetics present a high clinical potential to reduce the impact of contamination with blood and other biological fluids, which affects the performance of HCSCs [48], also allowing for final coronal restoration in the same clinical session. Further, a shortened setting time could potentially decrease the risk of cement wash-out by blood flow in procedures, such as root end surgery, which require short setting times of about 3–10 min [5]. While the bioactivity of ProRoot MTA has previously been confirmed, showing the capacity to promote cell proliferation and differentiation [9,49] and mineral forming ability from phosphate-containing saliva, the bioactivity of RS+ should be further studied.

Because of the favourable rheological and faster setting kinetics, synthetic, finer grained HCSCs, such as RS+, represent a viable option as an alternative to conventional MTA-based materials for endodontic treatments. However, the promising in vitro material performance (physio-chemical properties) of RS+ in terms of the cement composition, fineness, setting kinetics, and rheological properties should be complemented by a controlled, long-term clinical study in the aggressive environment of the oral cavity.

5. Conclusions

Distinct differences were observed between the benchmark MTA-based material ProRoot MTA and RS+, a next generation HCSC with a refined physio-chemical formulation. RS+ had a narrower monomodal PSD distribution with an order of magnitude-finer d_{mean} value compared to ProRoot MTA. SEM-EDA mapping confirmed the presence of homogeneously distributed calcium, silicon, and zirconium in the RS+, while in ProRoot MTA, in addition to calcium, silicon and bismuth, sulphur, aluminium, and iron were detected as well. The XRD analysis confirmed the much more diverse phase composition of ProRoot MTA.

While the kinetics of surface hydration of both HCSCs were comparable, it is concluded that the much finer biocompatible particulate formulation of RS+ proved pivotal in its ability to exert predictable viscous flow during the working time and more than two times faster viscoelastic-elastic transition, reflecting in improved handling and setting behaviour. Finally, RS+ could be completely transformed into hydration products within 48-h, while hydration products were not yet detected by XRD in ProRoot MTA since it bound to a surface thin film connected to the matrix of the unreacted particles.

Author Contributions: Conceptualisation, A.K. and A.P.J.; methodology, A.P.J., K.R., I.Ö., M.K. and A.K.; investigation, A.P.J. and K.R.; writing, A.P.J., T.M. and A.K.; supervision, A.K.; funding acquisition, A.K.; project administration, A.K.; validation, A.K.; visualisation, A.P.J. and T.M. All authors have read and agreed to the published version of the manuscript.

Funding: The Slovenian Research Agency is acknowledged for funding through research programs P2-0087, P2-0105, and P1-0045 and research projects J3-3064, N1-0189, and N1-0225.

Institutional Review Board Statement: Not applicable.

Informed Consent Statement: Not applicable.

Data Availability Statement: Not applicable.

Conflicts of Interest: The authors declare no conflict of interest.

References

1. Dawood, A.E.; Parashos, P.; Wong, R.H.K.; Reynolds, E.C.; Manton, D.J. Calcium silicate-based cements: Composition, properties, and clinical applications. *J. Investig. Clin. Dent.* **2017**, *8*, e12195. [CrossRef] [PubMed]
2. Simon, S.; Rilliard, F.; Berdal, A.; Machtou, P. The use of mineral trioxide aggregate in one-visit apexification treatment: A prospective study. *Int. Endod. J.* **2007**, *40*, 186–197. [CrossRef]
3. Gandolfi, M.G.; Siboni, F.; Botero, T.; Bossu, M.; Riccitiello, F.; Prati, C. Calcium silicate and calcium hydroxide materials for pulp capping: Biointeractivity, porosity, solubility and bioactivity of current formulations. *J. Appl. Biomater. Funct. Mater.* **2015**, *13*, 43–60. [CrossRef]
4. Hansen, S.W.; Marshall, J.G.; Sedgley, C.M. Comparison of intracanal EndoSequence Root Repair Material and ProRoot MTA to induce pH changes in simulated root resorption defects over 4 weeks in matched pairs of human teeth. *J. Endod.* **2011**, *37*, 502–506. [CrossRef]
5. Prati, C.; Gandolfi, M.G. Calcium silicate bioactive cements: Biological perspectives and clinical applications. *Dent. Mater.* **2015**, *31*, 351–370. [CrossRef]
6. Hashem, A.A.; Wanees Amin, S.A. The effect of acidity on dislodgment resistance of mineral trioxide aggregate and bioaggregate in furcation perforations: An in vitro comparative study. *J. Endod.* **2012**, *38*, 245–249. [CrossRef] [PubMed]
7. Torabinejad, M.; Chivian, N. Clinical applications of mineral trioxide aggregate. *J. Endod.* **1999**, *25*, 197–205. [CrossRef]
8. Nair, P.N.; Duncan, H.F.; Pitt Ford, T.R.; Luder, H.U. Histological, ultrastructural and quantitative investigations on the response of healthy human pulps to experimental capping with mineral trioxide aggregate: A randomized controlled trial. *Int. Endod. J.* **2008**, *41*, 128–150.
9. Paranjpe, A.; Zhang, H.; Johnson, J.D. Effects of mineral trioxide aggregate on human dental pulp cells after pulp-capping procedures. *J. Endod.* **2010**, *36*, 1042–1047. [CrossRef]
10. Mente, J.; Geletneky, B.; Ohle, M.; Koch, M.J.; Friedrich Ding, P.G.; Wolff, D.; Dreyhaupt, J.; Martin, N.; Staehle, H.J.; Pfefferle, T. Mineral trioxide aggregate or calcium hydroxide direct pulp capping: An analysis of the clinical treatment outcome. *J. Endod.* **2010**, *36*, 806–813. [CrossRef] [PubMed]
11. Chong, B.S.; Pitt Ford, T.R.; Hudson, M.B. A prospective clinical study of Mineral Trioxide Aggregate and IRM when used as root-end filling materials in endodontic surgery. *Int. Endod. J.* **2009**, *42*, 414–420. [CrossRef] [PubMed]
12. Pontius, V.; Pontius, O.; Braun, A.; Frankenberger, R.; Roggendorf, M.J. Retrospective evaluation of perforation repairs in 6 private practices. *J. Endod.* **2013**, *39*, 1346–1358. [CrossRef]
13. Mente, J.; Leo, M.; Panagidis, D.; Ohle, M.; Schneider, S.; Lorenzo Bermejo, J.; Pfefferle, T. Treatment outcome of mineral trioxide aggregate in open apex teeth. *J. Endod.* **2013**, *39*, 20–26. [CrossRef]
14. Gandolfi, M.G.; Van Landuyt, K.; Taddei, P.; Modena, E.; Van Meerbeek, B.; Prati, C. Environmental scanning electron microscopy connected with energy dispersive x-ray analysis and Raman techniques to study ProRoot mineral trioxide aggregate and calcium silicate cements in wet conditions and in real time. *J. Endod.* **2010**, *36*, 851–857. [CrossRef] [PubMed]
15. Richardson, I.G. Tobermorite/jennite- and tobermorite/calcium hydroxide-based models for the structure of C-S-H: Applicability to hardened pastes of tricalcium silicate, β-dicalcium silicate, Portland cement, and blends of Portland cement with blast-furnace slag, metakaolin, or silica fume. *Cem. Concr. Res.* **2004**, *34*, 1733–1777.
16. Dammaschke, T.; Gerth, H.U.; Zuchner, H.; Schafer, E. Chemical and physical surface and bulk material characterization of white ProRoot MTA and two Portland cements. *Dent. Mater.* **2005**, *21*, 731–738. [CrossRef]
17. Torabinejad, M.; Hong, C.U.; McDonald, F.; Pitt Ford, T.R. Physical and chemical properties of a new root-end filling material. *J. Endod.* **1995**, *21*, 349–353. [CrossRef]
18. Gandolfi, M.G.; Taddei, P.; Siboni, F.; Modena, E.; Ciapetti, G.; Prati, C. Development of the foremost light-curable calcium-silicate MTA cement as root-end in oral surgery. Chemical-physical properties, bioactivity and biological behavior. *Dent. Mater.* **2011**, *27*, e134–e157. [CrossRef] [PubMed]
19. Storm, B.; Eichmiller, F.C.; Tordik, P.A.; Goodell, G.G. Setting expansion of gray and white mineral trioxide aggregate and Portland cement. *J. Endod.* **2008**, *34*, 80–82. [CrossRef]
20. Boutsioukis, C.; Noula, G.; Lambrianidis, T. Ex vivo study of the efficiency of two techniques for the removal of mineral trioxide aggregate used as a root canal filling material. *J. Endod.* **2008**, *34*, 1239–1242. [CrossRef]
21. Lenherr, P.; Allgayer, N.; Weiger, R.; Filippi, A.; Attin, T.; Krastl, G. Tooth discoloration induced by endodontic materials: A laboratory study. *Int. Endod. J.* **2012**, *45*, 942–949. [CrossRef] [PubMed]
22. Felman, D.; Parashos, P. Coronal tooth discoloration and white mineral trioxide aggregate. *J. Endod.* **2013**, *39*, 484–487. [CrossRef]
23. Monteiro Bramante, C.; Demarchi, A.C.; de Moraes, I.G.; Bernadineli, N.; Garcia, R.B.; Spangberg, L.S.; Duarte, M.A. Presence of arsenic in different types of MTA and white and gray Portland cement. *Oral Surg. Oral Med. Oral Pathol. Oral Radiol. Endod.* **2008**, *106*, 909–913. [CrossRef]
24. Gandolfi, M.G.; Ciapetti, G.; Taddei, P.; Perut, F.; Tinti, A.; Cardoso, M.V.; Van Meerbeek, B.; Prati, C. Apatite formation on bioactive calcium-silicate cements for dentistry affects surface topography and human marrow stromal cells proliferation. *Dent. Mater.* **2010**, *26*, 974–992. [CrossRef] [PubMed]
25. Camilleri, J.; Montesin, F.E.; Brady, K.; Sweeney, R.; Curtis, R.V.; Ford, T.R. The constitution of mineral trioxide aggregate. *Dent. Mater.* **2005**, *21*, 297–303. [CrossRef]

26. Darvell, B.W.; Wu, R.C. "MTA"-an Hydraulic Silicate Cement: Review update and setting reaction. *Dent. Mater.* **2011**, *27*, 407–422. [CrossRef] [PubMed]
27. Setbon, H.M.; Devaux, J.; Iserentant, A.; Leloup, G.; Leprince, J.G. Influence of composition on setting kinetics of new injectable and/or fast setting tricalcium silicate cements. *Dent. Mater.* **2014**, *30*, 1291–1303. [CrossRef]
28. Koubi, G.; Colon, P.; Franquin, J.C.; Hartmann, A.; Richard, G.; Faure, M.O.; Lambert, G. Clinical evaluation of the performance and safety of a new dentine substitute, Biodentine, in the restoration of posterior teeth-a prospective study. *Clin. Oral Investig.* **2013**, *17*, 243–249. [CrossRef]
29. Grech, L.; Mallia, B.; Camilleri, J. Characterization of set Intermediate Restorative Material, Biodentine, Bioaggregate and a prototype calcium silicate cement for use as root-end filling materials. *Int. Endod. J.* **2013**, *46*, 632–641. [CrossRef]
30. Tanalp, J.; Karapinar-Kazandag, M.; Dolekoglu, S.; Kayahan, M.B. Comparison of the radiopacities of different root-end filling and repair materials. *Sci. World J.* **2013**, *2013*, 594950. [CrossRef]
31. Drnovsek, N.; Novak, S.; Dragin, U.; Ceh, M.; Gorensek, M.; Gradisar, M. Bioactive glass enhances bone ingrowth into the porous titanium coating on orthopaedic implants. *Int. Orthop.* **2012**, *36*, 1739–1745. [CrossRef]
32. Gantar, A.; da Silva, L.P.; Oliveira, J.M.; Marques, A.P.; Correlo, V.M.; Novak, S.; Reis, R.L. Nanoparticulate bioactive-glass-reinforced gellan-gum hydrogels for bone-tissue engineering. *Mater. Sci. Eng. C Mater. Biol. Appl.* **2014**, *43*, 27–36. [CrossRef]
33. Ha, W.N.; Nicholson, T.; Kahler, B.; Walsh, L.J. Methodologies for measuring the setting times of mineral trioxide aggregate and Portland cement products used in dentistry. *Acta Biomater. Odontol. Scand.* **2016**, *2*, 25–30. [CrossRef] [PubMed]
34. ISO 9917-1:2007; Dentistry—Water-Based Cements–Part 1: Powder/Liquid Acid-Base Cements. ISO: Geneva, Switzerland, 2007.
35. Shen, Y.; Peng, B.; Yang, Y.; Ma, J.; Haapasalo, M. What do different tests tell about the mechanical and biological properties of bioceramic materials? *Endod. Top.* **2015**, *32*, 47–85. [CrossRef]
36. Higl, J.; Hinder, D.; Rathgeber, C.; Ramming, B.; Lindén, M. Detailed in situ ATR-FTIR spectroscopy study of the early stages of C-S-H formation during hydration of monoclinic C3S. *Cem. Concr. Res.* **2021**, *142*, 106367. [CrossRef]
37. Black, L.; Breen, C.; Yarwood, J.; Garbev, K.; Stemmermann, P.; Gasharova, B. Structural Features of C–S–H(I) and Its Carbonation in Air—A Raman Spectroscopic Study. Part II: Carbonated Phases. *J. Am. Ceram. Soc.* **2007**, *90*, 908–917. [CrossRef]
38. Garbev, K.; Stemmermann, P.; Black, L.; Breen, C.; Yarwood, J.; Gasharova, B. Structural Features of C–S–H(I) and Its Carbonation in Air—A Raman Spectroscopic Study. Part I: Fresh Phases. *J. Am. Ceram. Soc.* **2007**, *90*, 900–907. [CrossRef]
39. Li, Q.; Coleman, N.J. The hydration chemistry of ProRoot MTA. *Dent. Mater. J.* **2015**, *34*, 458–465. [CrossRef] [PubMed]
40. Gandolfi, M.G.; Taddei, P.; Tinti, A.; Prati, C. Apatite-forming ability (bioactivity) of ProRoot MTA. *Int. Endod. J.* **2010**, *43*, 917–929. [CrossRef]
41. Taddei, P.; Modena, E.; Tinti, A.; Siboni, F.; Prati, C.; Gandolfi, M. Effect of the fluoride content on the bioactivity of calcium silicate-based endodontic cements. *Ceram. Int.* **2013**, *40*, 4095–4107. [CrossRef]
42. Lewis, J.A. Colloidal Processing of Ceramics. *J. Am. Ceram. Soc.* **2000**, *83*, 2341–2359. [CrossRef]
43. Camilleri, J. Mineral trioxide aggregate: Present and future developments. *Endodontic Topics* **2015**, *32*, 31–46. [CrossRef]
44. Du, M.; Liu, J.; Clode, P.; Leong, Y.-K. Microstructure and rheology of bentonite slurries containing multiple-charge phosphate-based additives. *Appl. Clay Sci.* **2019**, *169*, 120–128. [CrossRef]
45. Luckham, P.F.; Rossi, S. The colloidal and rheological properties of bentonite suspensions. *Adv. Colloid Interface Sci.* **1999**, *82*, 43–92. [CrossRef]
46. Logeshwaran, A.; Elsen, R.; Nayak, S. Mechanical and biological characteristics of 3D fabricated clay mineral and bioceramic composite scaffold for bone tissue applications. *J. Mech. Behav. Biomed. Mater.* **2023**, *138*, 105633. [CrossRef] [PubMed]
47. Zhou, H.M.; Shen, Y.; Zheng, W.; Li, L.; Zheng, Y.F.; Haapasalo, M. Physical properties of 5 root canal sealers. *J. Endod.* **2013**, *39*, 1281–1286. [CrossRef]
48. Gandolfi, M.G.; Iacono, F.; Agee, K.; Siboni, F.; Tay, F.; Pashley, D.H.; Prati, C. Setting time and expansion in different soaking media of experimental accelerated calcium-silicate cements and ProRoot MTA. *Oral Surg. Oral Med. Oral Pathol. Oral Radiol. Endod.* **2009**, *108*, e39–e45. [CrossRef]
49. Takita, T.; Hayashi, M.; Takeichi, O.; Ogiso, B.; Suzuki, N.; Otsuka, K.; Ito, K. Effect of mineral trioxide aggregate on proliferation of cultured human dental pulp cells. *Int. Endod. J.* **2006**, *39*, 415–422. [CrossRef] [PubMed]

Disclaimer/Publisher's Note: The statements, opinions and data contained in all publications are solely those of the individual author(s) and contributor(s) and not of MDPI and/or the editor(s). MDPI and/or the editor(s) disclaim responsibility for any injury to people or property resulting from any ideas, methods, instructions or products referred to in the content.

Article

Three-Dimensional Printed Teeth in Endodontics: A New Protocol for Microcomputed Tomography Studies

Tiago Reis [1,2,3,*], Cláudia Barbosa [3,4], Margarida Franco [2], Ruben Silva [2], Nuno Alves [2], Pablo Castelo-Baz [5], Jose Martín-Cruces [1] and Benjamín Martín-Biedma [5]

1. Endodontics and Restorative Dentistry Unit, School of Medicine and Dentistry, University of Santiago de Compostela, 15701 Santiago de Compostela, Spain; pepe3214@gmail.com
2. Centre for Rapid and Sustainable Product Development (CDRSP), Polytechnic University of Leiria, 2411-901 Leiria, Portugal; margarida.franco@ipleiria.pt (M.F.); ruben.j.silva@ipleiria.pt (R.S.); nuno.alves@ipleiria.pt (N.A.)
3. FP-I3ID, FP-BHS, Health Sciences Faculty, University Fernando Pessoa, 4249-004 Porto, Portugal; cbarbosa@ufp.edu.pt
4. RISE-Health, University Fernando Pessoa, 4249-004 Porto, Portugal
5. Oral Sciences Research Group, Endodontics and Restorative Dentistry Unit, School of Medicine and Dentistry, University of Santiago de Compostela, Health Research Institute of Santiago de Compostela (IDIS), 15706 Santiago de Compostela, Spain; pablocastelobaz@hotmail.com (P.C.-B.); benjamin.martin@usc.es (B.M.-B.)
* Correspondence: tiagofaria@ufp.edu.pt

Citation: Reis, T.; Barbosa, C.; Franco, M.; Silva, R.; Alves, N.; Castelo-Baz, P.; Martín-Cruces, J.; Martín-Biedma, B. Three-Dimensional Printed Teeth in Endodontics: A New Protocol for Microcomputed Tomography Studies. *Materials* **2024**, *17*, 1899. https://doi.org/10.3390/ma17081899

Academic Editors: Luigi Generali, Vittorio Checchi and Eugenio Pedullà

Received: 9 March 2024
Revised: 18 April 2024
Accepted: 18 April 2024
Published: 19 April 2024

Copyright: © 2024 by the authors. Licensee MDPI, Basel, Switzerland. This article is an open access article distributed under the terms and conditions of the Creative Commons Attribution (CC BY) license (https:// creativecommons.org/licenses/by/ 4.0/).

Abstract: This study aimed to describe a support material removal protocol (SMRP) from inside the root canals of three-dimensional printed teeth (3DPT) obtained by the microcomputed tomography (microCT) of a natural tooth (NT), evaluate its effectiveness by comparing the 3DPT to NT in terms of internal anatomy and behaviour toward endodontic preparation, and evaluate if 3DPT are adequate to assess the differences between two preparation systems. After the SMRP, twenty 3DPT printed by PolyJet™ were microCT scanned before preparation and thereafter randomly assigned into two groups (n = 10). One group and NT were prepared using ProTaper Gold® (PTG), and the other group with Endogal® (ENDG). MicroCT scans were carried out after preparation, and the volume increase, volume of dentin removed, centroids, transportation, and unprepared areas were compared. For the parameters evaluated, no significant differences were found between the 3DPT and NT before and after preparation ($p > 0.05$), and no significant differences were found between the 3DPT PTG group and the 3DPT ENDG group ($p > 0.05$). It can be concluded that the SMRP described is effective in removing the support material SUP706B™. PolyJet™ is adequate for printing 3DPT. Furthermore, 3DPT printed with high-temperature RGD525™ have similar behaviour during endodontic preparation with PTG as the NT, and 3DPT can be used to compare two preparation systems.

Keywords: endodontics; three-dimensional printed teeth; PolyJet; support material; microcomputed tomography

1. Introduction

The American Association of Endodontists defines root canal preparation as "Procedures involved in cleaning and shaping the canal system prior to obturation", distinguishing between "biomechanical preparation" as the "use of rotary/reciprocating and/or hand instruments to expose, clean, enlarge and shape the pulp canal space, usually in conjunction with irrigants" and "chemomechanical preparation" as the "use of chemicals for irrigation of the root canal, demineralization of dentin, dissolution of pulp tissue and neutralization of bacterial products and toxins; used in conjunction with biomechanical preparation" [1].

Over the last few decades, root canal preparation protocols have changed, and many new nickel–titanium systems have become available; nonetheless, clinicians require knowledge of shaping properties and a performance evaluation of these systems to select them according to clinical cases [2]. Root canal preparation ex vivo studies provide useful and valuable data to improve the biological outcome of preparation and therefore have to be continued in the future [3].

Extracted human natural teeth (NT) are considered the gold standard for ex vivo studies; however, they present several disadvantages, the main one being very difficult standardization [4], not only as a consequence of the root canal system anatomy but also the donor's age, which has an influence on the dentin properties [5,6]. Three-dimensional printed teeth (3DPT) obtained through the microcomputed tomography (microCT) of real NT have been used as an alternative for NT in both research and teaching and offer a good opportunity to create balanced experimental groups [3,7–14]. The major concern expressed about 3DPT is the difference in radiopacity and hardness between resin and human dentin [3,7,9,10,12]. Nonetheless, protocols for the standardization of studies using 3DPT still have to be developed [3].

PolyJet® (Stratasys Ltd., Eden Prairie, MN, USA) printing is based on layer-by-layer technology. The process consists of the nozzles of the printer moving along the XY plane and spraying liquid photosensitive resin on the printer bed, and a UV lamp cures the resin. After the first layer is finished, the printer bed will drop by a layer thickness in the Z plane, and the deposition of another layer is repeated. This is a potentially attractive option for low-volume manufacturing in research environments. Where hollow parts or overhangs exist, the nozzles spray a layer of removable support material [15–17]. Support material removal methods include manual breaking, dissolution under water pressure or with sodium hydroxide (NaOH), or melting [15,18,19]. However, it has been reported that there is difficulty in removing the support material inside 3DPT root canals [3,7,8,17,20], so canals could be filled partially or totally with support material, which may have an effect on microCT analyses before and/or after preparation [21].

In order for 3DPT to be used in ex vivo preparation studies, their internal anatomy should be similar to the original NT, they should be support-material-free, their behaviour during preparation should be identical to that of NT, and they should be capable of assessing the differences between two preparation systems. To the authors' knowledge, there are no studies that describe a protocol for removing the support material from inside the root canals of 3DPT printed by PolyJet™. In addition, the available literature regarding the printing accuracy of internal anatomy is still scarce.

Therefore, the first aim of this study was to evaluate the effectiveness of a support material removal protocol (SMRP) for inside root canals by comparing the internal anatomy of a NT with 3DPT. The null hypothesis was that there was no difference between the internal anatomy of NT and 3DPT. The second aim of this study was to compare the behaviour under preparation with ProTaper Gold® (Dentsply-Sirona, Fair Lawn, NJ, USA) (PTG) between NT and 3DPT. The null hypothesis was that there was no difference in preparation behaviour between NT and 3DPT. The third aim of this study was to compare the behaviour of 3DPT during preparation with two different preparation systems, namely PTG and Endogal® (Endogal, Galician Endodontics Company, Lugo, Spain) (ENDG). The null hypothesis was that no difference existed between PTG and ENDG.

2. Materials and Methods

The study protocol was approved by the ethics committee of Fernando Pessoa University (FCS/PI 429/23). Based on data from a previous study [22] in which shaping ability was assessed, a sample size calculation was performed using G*Power 3.1.9.7 software for Windows (Heinrich Heine, Universität Düsseldorf, Düsseldorf, Germany) with an α type error of 0.05 and a β power of 0.95 for an effect size of 1.79 input into the t test family, resulting in a required sample size of 16 samples (8 per group) to observe significant differences between groups. Ten samples were used per group to compensate for possible sample loss during experimental procedures.

2.1. Natural Specimen Selection

An initial pool of 45 maxillary permanent molars, extracted for reasons unrelated to this study, was used. The teeth were collected and stored in distilled water until use. Radiographs were taken in mesiodistal and buccolingual directions to ensure that the inclusion criteria were met. The inclusion criteria were teeth with fully formed apices, the absence of root fractures, no signs of external and internal resorption or decayed tissue in the region of interest, the absence of previous endodontic treatment, and a degree of curvature between 20° and 40°. The degree of curvature was measured according to Schneider's method [23]. From the initial sample, 14 teeth were selected. The endodontic cavities were prepared using a round diamond bur #4 and an Endo-Z™ bur (Dentsply Sirona, Fair Lawn, NJ, USA) driven by a high-speed handpiece under water cooling. The canals were explored with a K-file #10 (Dentsply Sirona) until the tip of the file was just visible through the apical foramen to ensure the existence of canal patency. The tooth crowns were sectioned at 4 mm from the cementoenamel junction to create a platform for easing future references. The specimens were then scanned using a microCT device (Skyscan 1174; Bruker, Kontich, Belgium) at 50 KV and 800 mA energy; a 0.25 mm thick aluminium filter was used, with rotational steps of 1.0 increments for a total rotation of 180°, a 16.65 µm image pixel size, and an exposure time of 12,000 ms. Images were reconstructed using NRecon version 1.7.46 software (Bruker, Kontich, Belgium, in which algorithms were introduced for the correction of ring artefacts (3), smoothing (3), and beam hardening (40%). CTAn version 1.20.3.0 software (Bruker, Kontich, Belgium) was applied to produce one STL file of each tooth and another of the respective canal's anatomy. The STL files were exported to a free software platform (MeshLab version 2021.10) for qualitative canal configuration evaluation and the election of the NT to replicate. The tooth chosen was a second left maxillary molar that presented 3 fully separated roots, each one with a single independent canal, without any lateral canals. The mesiobuccal root presented an oval canal with a degree of curvature of 32°; the distobuccal root presented a round small diameter canal with a degree of curvature of 30°; and the palatal root presented a round large diameter canal with a degree of curvature of 26° (Figure 1).

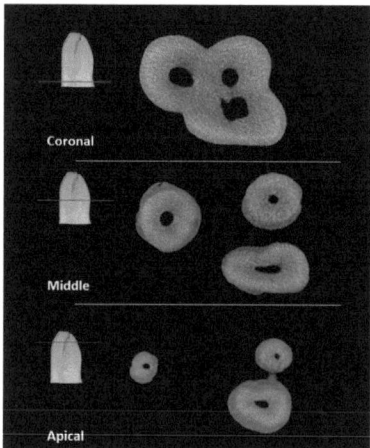

Figure 1. Cross-sectional views of microCT scan of the natural tooth.

The STL file of the tooth was simplified and prepared for 3D printing using the "Simplification: Quadric Edge Collapse Decimation" filter [14] because a high-resolution STL file is an excessively large file that 3D printing softwares has difficulties processing [4]. The original STL file presented 3,298,498 vertices and 6,598,466 faces with a 314 Mbytes size, and the simplified STL presented 798,881 vertices and 1,599,232 faces with a 76.2 Mbytes size. The STL file was printed by PolyJet™ using a Stratasys Object30 Prime™ printer (Stratasys Ltd., Eden Prairie, MN, USA) in high-quality mode with a layer thickness of 16 µm. High-temperature RGD525™ (Stratasys Ltd.) and the support material SUP706B™ (Stratasys Ltd.) were the materials chosen for printing. The tooth was displayed horizontally with the mesial surface in contact with the printer bed, with its long axis parallel to the X axis of the printer head and printer bed (Figure 2).

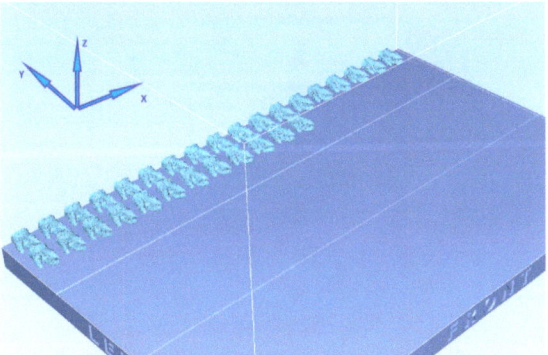

Figure 2. Teeth orientation on the printer bed.

After printing, the support material involving the teeth and inside the access cavity was removed manually. The support material removal protocol (SMRP), summarized in a flowchart (Figure 3), involved using a K-file #15 (Dentsply Sirona), a 30G polypropylene body needle Irriflex® (Produits dentaires SA, Vevey, Switzerland), and an Endoactivator® (Dentsply Sirona) with a small Endoactivator® Tip (15/0.02) (Dentsply Sirona). Irrigants were used, namely a 5% solution of NaOH, 5% Derquim® LM 01 alkalin detergent (ITW Reagents, S.R.L., Castellar del Vallès, Spain), distilled water, and 70% alcohol. The SMRP steps were (1) access cavity cleaning with 5 mL of 5% NaOH, and for each canal, (2) advance passively into the canal the K-file #15 until the tip of the file is just visible through the apical foramen to ensure the patency of the canal, followed by irrigation with 5 mL of 5% NaOH at high pressure. This step was repeated 5 times to achieve irrigant extrusion at the end of this step. Then, it was followed by (3) irrigation with 35 mL of 5% NaOH; during this step, if irrigant extrusion was lost, meaning losing the patency, a K-file #15 was used as before; (4) irrigant sonic activation with an Endoactivator® (Dentsply Sirona) at a high frequency for 30 s with up-and-down movements, with an amplitude of 4 mm; (5) irrigation with 5 mL of 5% NaOH; (6) irrigant sonic activation as described before; (7) irrigation with 5 mL of 5% Derquim® LM 01 alkalin detergent; (8) irrigant sonic activation as described before; (9) irrigation with 5 mL of distilled water; (10) irrigant sonic activation as described before; (11) irrigation with 5 mL of distilled water; and (12) irrigation with 2 mL of 70% alcohol.

In this way, a 20-3DPT sample (n = 20) randomly assigned (www.random.org, accessed on 10 January 2023) to each of the two experimental groups (n = 10), PTG and ENDG (Figure 4), was obtained.

In sequence, the 3DPT were scanned using a microCT device (Skyscan 1174; Bruker) at 50 KV; 800 mA of energy; and rotational steps of 1.0 increments for a total rotation of 180° with a 16.65 µm image pixel size and an exposure time of 12,000 ms. The images were reconstructed using NRecon v 1.7.46 software, in which algorithms were introduced for the correction of ring artefacts (5), smoothing (3), and beam hardening (50%).

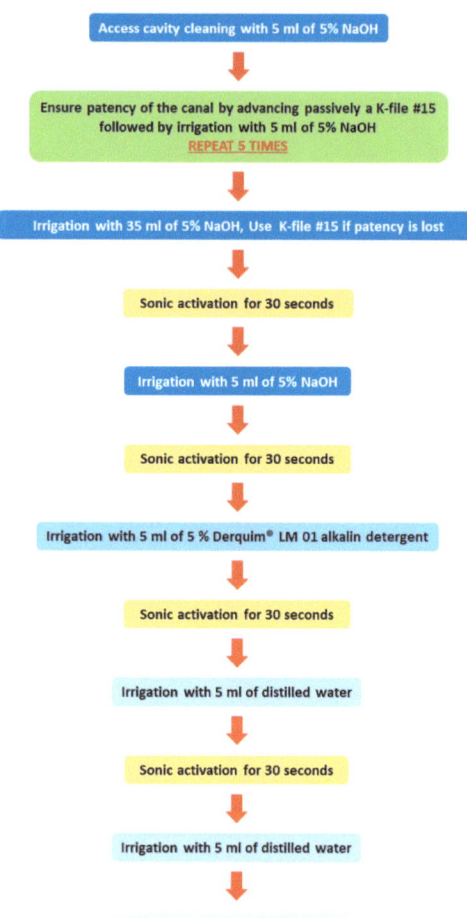

Figure 3. Flowchart of support material removal protocol.

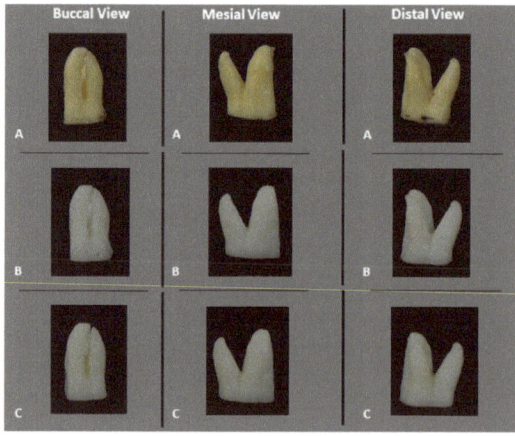

Figure 4. Natural teeth (**A**) and 3D-printed teeth (**B**,**C**) from different views.

2.2. Root Canal Preparation

All the preparations were performed by a single operator with 22 years of clinical experience in the field of endodontics and previous experience in using PTG and ENDG systems clinically. The teeth were mounted in place using the ProTrain system® (Simit Dental Srl, Mantua, Italy). The working length (WL) was determined by taking 1 mm from the value obtained during the SMRP. The WL was 16 mm for the buccal canals and 15 mm for the palatal canal. An electric motor X-Smart® Plus (Dentsply Sirona) was used to operate the files with in-and-out pecking motion (2–3 mm amplitude) in a continuous clockwise rotation according to the manufacturer's recommendations.

2.3. Preparation of NT and Group I with PTG

A glide path was created by using a ProGlider® instrument (Dentsply Sirona) (a size tip of 16 and progressive taper from 0.02 to 0.08) until the WL was reached. All files from the PTG system, which have a convex triangular cross section and progressive taper [24], were used up to the WL except the SX file, which was used only for coronal interference removal, in the sequence SX (19/0.04), S1 (18/0.02), S2 (20/0.04), F1 (20/0.07), and F2 (25/0.08). Patency was checked after the use of each instrument with K-file #10. The instruments were used for one tooth preparation; after that, they were discharged. Root canal irrigation was performed between each file with 5 mL of 5.25% NaOCl (Cerkamed, Stalowa Wola, Poland) for NT and 5 mL of distilled water for 3DPT using an Irriflex® (Produits dentaires SA, Vevey, Switzerland) needle and activated with an Endoactivator® at a high frequency with a small tip for 30 s. After preparation, the canals were irrigated two times with 5 mL of irrigant and activated with the Endoactivator® (Dentsply Sirona), so the total volume of irrigation for each canal was 40 mL. In the end, the canals were dried with paper points (Dentsply Sirona).

2.4. Preparation of Group II with ENDG

A glide path was created by using the A (15/0.03) instrument until the WL was reached. All files from the ENDG system, which is a new system with a parallelogram cross section and instruments of constant 4% and 6% taper [25], were used up to the WL, except the X file, which was used only for coronal interference removal, in the sequence X (25/0.09), B (20/0.04), C (25/0.04), and D (25/0.06). Patency was checked after the use of each instrument with K-file #10. The instruments were used for one tooth preparation; after that, they were discharged. Root canal irrigation was performed between each file with 5 ml of distilled water using an Irriflex® needle and activated with the Endoactivator® at a high frequency with a small tip for 30 s. After preparation, the canals were irrigated three times with 5 mL of irrigant and activated with Endoactivator®, so the total volume of irrigation for each canal was 40 mL. In the end, the canals were dried with paper points.

A new microCT scan was performed on NT and both 3DPT groups according to the same scanning and reconstruction parameters as those established initially, respectively.

2.5. MicroCT Evaluation

A microCT evaluation was conducted by one of the authors, blinded to the groups. Images before and after preparation were superimposed with the 3D registration application of the DataViewer v 1.5.6.2 software (Bruker microCT), and the data obtained were processed using CTAnv v 1.20.3.0 software (Bruker microCT). The region of interest was set from the furcation region to the apex of the root.

The volume of dentine removed and centroids were quantified by subtracting the values before preparation from the values after preparation [3]. In accordance with the orientation in which the samples were evaluated, a positive value for the alteration of centroid X meant an alteration in the buccal direction, centroid Y in the mesial direction, and centroid Z in the apical direction [26]. The transportation was investigated by calculating the vectorial translocation of all sections' X, Y, and Z coordinate values using the following formula, where "a" is after and "b" is before preparation: $\sqrt{(X_a - X_b)^2 + (Y_a - Y_b)^2 + (Z_a - Z_b)^2}$ [27].

The percentage of unprepared area was calculated by the number of static voxels compared with the total number of voxels present on the root canal surface [26]. The tooth volume expansion evaluated for the 3DPT corresponds to the ratio of the 3DPT tooth volume compared to the NT volume [13].

2.6. Statistical Analysis

The collected data were processed using IBM SPSS Statistics version 26.0 software. The Shapiro–Wilk test was applied to verify data normality. Accordingly, for normal and non-normal distributions, a one-sample t-test or one-sample Wilcoxon signed-rank test and an independent-sample t-test or Mann–Whitney U test were applied. The significance level was 5% for all statistical tests ($p < 0.05$).

3. Results

The measurements of canal volume, centroid X, Y, and Z before preparation, and the percentage of tooth volume expansion for the 3DPT are shown in Table 1. There were no statistically significant differences between the NT and the 3DPT for all the variables ($p > 0.05$) (Figures 5 and 6).

Table 1. Microcomputed tomographic analysis before preparation of natural teeth and 3D-printed teeth.

Data	Natural Tooth	3D-Printed Teeth	
Canal Volume (mm^3)	12.08	Mean ± SD	11.99 ± 0.55
		Median	12.07
Centroid X (mm)	10.92	Mean ± SD	10.92 ± 0.09
		Median	10.93
Centroid Y (mm)	10.37	Mean ± SD	10.35 ± 0.07
		Median	10.35
Centroid Z (mm)	8.03	Mean ± SD	8.02 ± 0.09
		Median	8.04
Tooth Volume Expansion (%)		Mean ± SD	0.73 ± 0.77
		Median	0.68

A comparison between the NT and 3DPT before and after preparation with the PTG system is presented in Table 2. There were no statistically significant differences between the NT and the 3DPT for all the variables before preparation ($p > 0.05$). The 3DPT showed a similar behaviour to the NT after preparation since there were no statistically significant differences ($p > 0.05$) related to the measurements of canal volume, centroid X, Y, and Z after preparation, the volume of dentin removed, the percentage of volume increase, centroid X, Y, and Z alteration, transportation, and the percentage of unprepared areas (Figures 6 and 7).

The overall results of the measurements before preparation, as well as the results of the shaping performance of the PTG and ENDG instrument systems on the 3DPT, are presented in Table 3. There were no statistically significant differences between the two groups for all the variables before and after preparation ($p > 0.05$). (Figures 6 and 7).

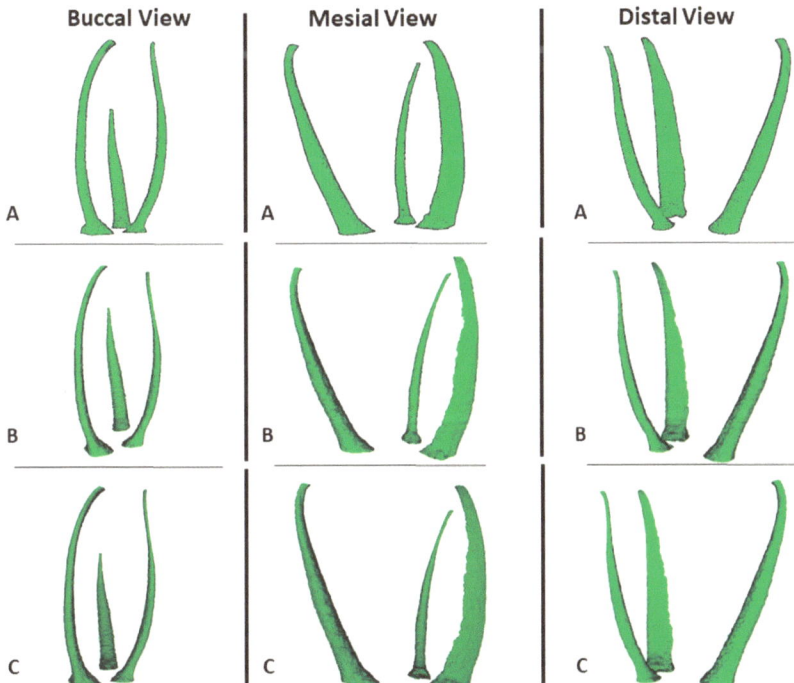

Figure 5. Representative 3D reconstruction of microCT scans after the support material removal protocol and before preparation from different views. (**A**) natural tooth; (**B**) 3D-printed teeth by ProTaper Gold® group; and (**C**) 3D-printed teeth by Endogal® group.

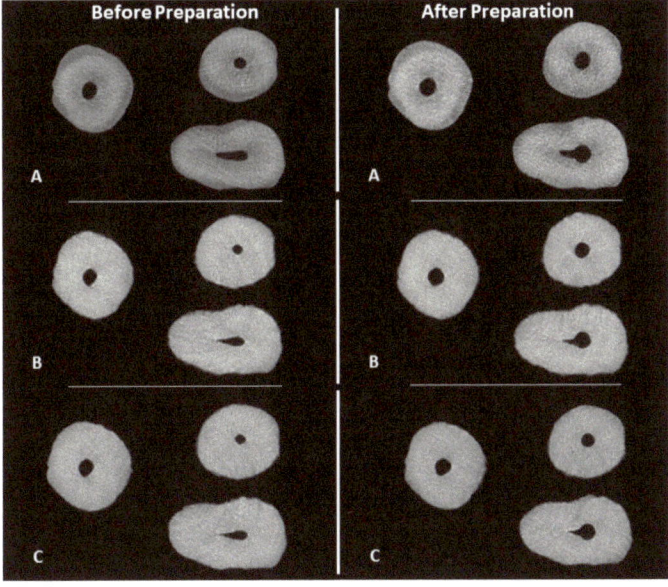

Figure 6. Cross-sectional views of microCT scans before and after preparation. (**A**) natural tooth; (**B**) 3D-printed teeth with ProTaper Gold® group; and (**C**) 3D-printed teeth with Endogal® group.

Table 2. Microcomputed tomographic analysis before and after preparation with ProTaper Gold® of natural tooth and 3D-printed teeth.

Data			Natural Tooth	3D-Printed Teeth	
Canal Volume (mm³)	Initial		12.08	Mean ± SD	12.12 ± 0.42
				Median	12.12
	After		17.75	Mean ± SD	17.89 ± 0.69
				Median	17.66
	Volume of dentin removed		5.68	Mean ± SD	5.77 ± 0.66
				Median	5.67
	% Volume increase		46.99	Mean ± SD	47.68 ± 6.09
				Median	45.50
Centroid X (mm)	Initial		10.92	Mean ± SD	10.92 ± 0.07
				Median	10.87
	After		11.38	Mean ± SD	11.33 ± 0.13
				Median	11.32
	Alteration		0.46	Mean ± SD	0.41 ± 0.17
				Median	0.41
Centroid Y (mm)	Initial		10.37	Mean ± SD	10.37 ± 0.07
				Median	10.40
	After		10.50	Mean ± SD	10.53 ± 0.04
				Median	10.53
	Alteration		0.13	Mean ± SD	0.16 ± 0.07
				Median	0.15
Centroid Z (mm)	Initial		8.03	Mean ± SD	8.01 ± 0.11
				Median	8.04
	After		8.10	Mean ± SD	8.01 ± 0.10
				Median	8.02
	Alteration		0.07	Mean ± SD	0.01 ± 0.015
				Median	0.02
Transportation (mm)			0.48	Mean ± SD	0.48 ± 0.14
				Median	0.45
Unprepared area (%)			57.64	Mean ± SD	54.97 ± 3.79
				Median	53.88

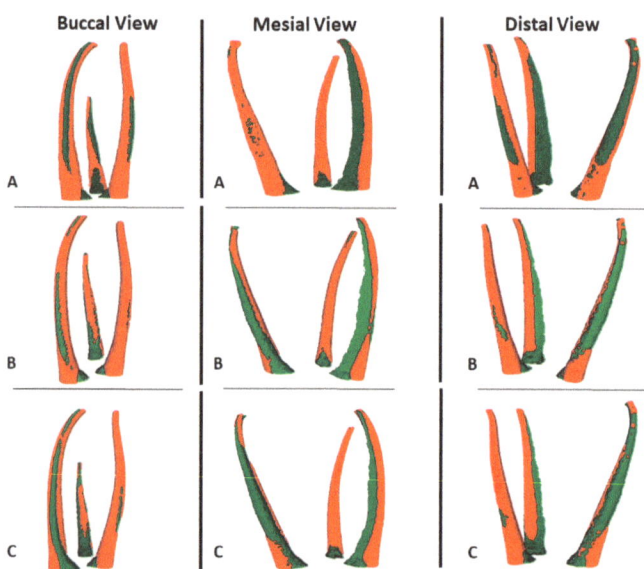

Figure 7. Representative 3D reconstruction of microCT scans before (green) and after (red) preparation from different views. (**A**) natural tooth; (**B**) 3D-printed teeth with ProTaper Gold® group; and (**C**) 3D-printed teeth with Endogal® group.

Table 3. Microcomputed tomographic analysis before and after preparation with ProTaper Gold® and EndoGal® of 3D-printed teeth.

Data			ProTaper Gold®	EndoGal®
Canal Volume (mm³)	Initial	Mean ± SD	12.12 ± 0.42	11.86 ± 0.66
		Median	12.12	11.82
	After	Mean ± SD	17.89 ± 0.69	17.02 ± 0.61
		Median	17.66	17.06
	Volume of dentin removed	Mean ± SD	5.77 ± 0.66	5.24 ± 1.02
		Median	5.67	5.49
	% Volume increase	Mean ± SD	47.68 ± 6.09	45.02 ± 10.89
		Median	45.50	47.25
Centroid X (mm)	Initial	Mean ± SD	10.92 ± 0.07	10.92 ± 0.11
		Median	10.87	10.96
	After	Mean ± SD	11.33 ± 0.13	11.37 ± 0.06
		Median	11.32	11.35
	Alteration	Mean ± SD	0.41 ± 0.17	0.45 ± 0.09
		Median	0.41	0.43
Centroid Y (mm)	Initial	Mean ± SD	10.37 ± 0.07	10.33 ± 0.06
		Median	10.40	10.32
	After	Mean ± SD	10.53 ± 0.04	10.44 ± 0.05
		Median	10.53	10.45
	Alteration	Mean ± SD	0.16 ± 0.07	0.11 ± 0.05
		Median	0.15	0.13
Centroid Z (mm)	Initial	Mean ± SD	8.01 ± 0.11	8.03 ± 0.07
		Median	8.04	8.01
	After	Mean ± SD	8.01 ± 0.10	8.07 ± 0.09
		Median	8.02	8.11
	Alteration	Mean ± SD	0.01 ± 0.015	0.05 ± 0.09
		Median	0.02	0.02
Transportation (mm)		Mean ± SD	0.48 ± 0.14	0.49 ± 0.11
		Median	0.45	0.47
Unprepared area (%)		Mean ± SD	54.97 ± 3.79	56.41 ± 5.11
		Median	53.88	55.39

4. Discussion

The present study described an SMRP and compared the root canal anatomy between a NT and 3DPT based on a microCT scan of the NT sample. Comparing the volume and centroids X, Y, and Z of the NT and the 3DPT before preparation, our findings demonstrated no statistically significant differences between them. Therefore, the null hypothesis was accepted. These findings show that the SMRP described is effective in removing the support material SUP706B™ used in this study, and PolyJet™ is adequate for printing 3DPT with similar internal anatomy to NT. The SMRP described here is based on the generic manufacturers' indications for removing the support material SUP706B™ by solubilization, in which an alkaline 2% solution of NaOH and a 1% solution of sodium metasilicate are used in a cleaning station, followed by water rinsing [19]. NaOH is the basis of the SMRP described here, and Derquim® LM 01 alkalin detergent was used, which is a detergent based on NaOH and anionic and non-ionic surfactants, to exert cleaning action on the root canal walls to mimic the detergent function of sodium metasilicate.

Sonic activation with an Endoactivator® was used to improve the irrigants' action by producing intra-canal irrigant agitation and streaming. Since the contact of the Endoactivator® tip with the root canal walls inhibits its free oscillation, reducing irrigant streaming, the tip was placed as deeply as possible without contacting the walls [28,29]. Nevertheless, even if contact occurs, the Endoactivator® tip is a polyamide, which does not produce active root canal cutting, diminishing the risk of root canal anatomy alteration [30]. The Irriflex® needle used is a polyethylene flexible needle that has a smoother progression inside the canal and does not wedge against root canal walls. It presents two side openings that produce two jets oriented in the direction of the root canal walls and delivers a large volume of irrigant at a high flow rate with a clinically minimal risk of apical extrusion [31]. However, for the SMRP to be effective, the irrigation must be carried out at high pressure and with the needle tip as close to the apex as possible for irrigant extrusion to occur.

Nevertheless, our findings show differences in the values of the 3DPT before the instrumentation of the parameters evaluated, meaning that the 3DPT are not equal between them. Relative to the root canal anatomy, this could be a result of our protocol since canal patency is mandatory for its effectiveness, being traduced, as stated before, in the visual observation of the extrusion of the irrigants. K-file #15 was chosen for this purpose since in channels under 150 μm, it is very difficult to remove the support material [17], and it was used obligatorily five times for each canal. After this number of utilizations, if the patency was lost, K-file #15 was used to reestablish it. Although used in a passive way, this means that the number of K-file #15 insertions was not the same for all the canals, and its effects on the root canal anatomy should be considered [20,32].

In relation to the total volume of the 3DPT, the alterations in volume that occur with PolyJet™ printing in order to expand have been described [33,34]. The findings of this study show that there is a 0.77% volume expansion, and this can also explain the differences between the 3DPT. The result of the present study is comparable to the 0.71% referred to in a previous study [13]; however, in that study, the STL file of the 3DPT and the volumetric analyses were realised by a 3D scanner in contrast with the microCT methodology of the present study.

Another issue that should be addressed is that, although PolyJet™ has the smallest dimensional error compared to other 3D printing technologies [16,35–37], its accuracy depends on the material used, the geometry of the printed object, and the orientation given in relation to both the printer head and the printer bed [15,34,38]. In a previous study, a rectangular flat part was printed with the long axis parallel to the X, Y, or Z axis of the printer bed, and it was concluded that the long axis of the printed part should be parallel to the X axis [38]. However, to the authors' knowledge, this feature was never studied while printing teeth, and in a single root tooth, the long axis would have been easily defined as the long axis of the root. In the present study, a multiple-root tooth was used, so it was measured as a whole, and the longest distance was in the buccal-to-palatal direction, so this was considered the long axis, and the 3DPT were printed as shown in Figure 2. Nevertheless, future studies should assess which is the best orientation to produce a multiple-root tooth according to the X, Y, or Z axis of the printer bed and if this has relevance to the accuracy of 3DPT. Also, it should be noted that the STL file was simplified and prepared for 3D printing, and this action may also have resulted in some level of distortion [13]. Future studies should assess which level of simplification is supported by the STL file without significant distortion relative to NT.

The NT used in the present study had its crown sectioned and presented an internal anatomy with three single different types of canals; however, it did not present any anatomical irregularities such as lateral canals or isthmus, which are known to be only accessible to irrigants [39]. Future studies are needed to evaluate the effectiveness of the support material in teeth with a full crown and with anatomical irregularities.

Regarding the second aim of this study, in order for 3DPT to be used in ex vivo preparation studies, their behaviour during preparation should be like NT. The present results regarding the volume of dentin removed, the percentage of volume increase, transportation, and unprepared areas show that there were no statistically significant differences between them. Therefore, the null hypothesis was accepted, and it can be concluded that 3DPT behave in a similar way to the NT when prepared with PTG. The major concern expressed about 3DPT is the difference in hardness between resin and human dentin [3,7,9,10,12]. High-temperature RGD525™ is an opaque model material that has exceptional dimensional stability and presents a tensile strength of 70–80 Mpa, a modulus of elasticity of 3200–3500 Mpa, and a flexural strength of 110–130 MPa [40]. In comparison, human dentin presents a tensile strength of

44.4–97.8 MPa, with a lower value for the inner dentin near to the pulp [41], a modulus of elasticity of 1375–1931 MPa [42], and a flexural strength of 171–254 MPa [43]. All these values between the printing material and dentin are approximated; however, it should be noticed that the material structure differs from the tubular structure of dentin.

Regarding the third aim, the results for the volume of dentin removed, the percentage of volume increase, transportation, and unprepared areas show that there were no statistically significant differences between the two preparation systems. So, the null hypothesis was accepted. Nevertheless, although there are no statistically significant differences, it is observed that ENDG produces a lower volume increase, a lower volume of dentin removed, and a higher percentage of unprepared area, thereby demonstrating that 3DPT are suitable for evaluating the differences between two preparation systems.

To the authors' knowledge, there is no other study that evaluates, using microCT, the shaping properties of the ENDG system, so the present results cannot be compared to others. Regarding the results observed for the PTG system in the present study, the percentage of volume increase was 47.68%, and in the literature, it ranges from 18.7% to 163.32% [22,24,44–46]. The percentage of unprepared area was 54.97%, while in the literature, it ranges from 3.57% to 46.85% [22,24,44–48]. The differences between the present study results and the literature can be explained by the type of teeth or canals used. Most studies use mandibular molar mesial canals; these studies show higher values for the percentage of volume increase and smaller values for the percentage of unprepared area [22,24,45,46], compared to, for instance, studies that use, for example, mandibular incisors or premolars [44,48]. The variation in the canal geometry has an effect on the preparation techniques [49]. It is widely accepted that preparing oval-shaped canals is a challenge, and smaller values for the percentage of unprepared area in this type of canal are associated with brushing movements during preparation and not with pecking movements [50,51].

In the present study, all the preparations were carried out by an experienced dentist in the field of endodontics with clinical experience with the two systems, and the major critical comment of the operator was a higher screw-in effect in the 3DPT compared to the NT, which results in a more difficult preparation, even with a controlled pecking movement of a 2–3 mm amplitude. The screw-in effect is the tendency of a rotary instrument to be pulled into the canal. It is affected by the type of movement kinematics, the cross section, or the taper of the instrument, as well as by the rotational speed [52–55]. In the present study, the manufacturer's recommendations of rotation per minute (rpm) and torque were used. Future studies are needed to establish how changes in rpm, torque, pecking movement amplitude values, and instrument design can influence 3DPT preparation.

As said before, protocols for the standardization of studies using 3DPT still have to be developed [3]. The present study, within its limitations, presents a description of an effective protocol for support material SUP706B™ removal, demonstrating that 3DPT printed with high-temperature RGD525™ material have similar behaviour during endodontic preparation to that of NT. Nonetheless, we propose that future research needs to achieve a standardisation of studies using 3DPT. In summary, the establishment of the optimal orientation regarding the printer bed of multiple-root teeth; the maximal level of simplification of the STL file without losing information if the SMRP described is effective in other types of root morphology; and establishing how changes in the values of the rpm, torque, and amplitude of pecking movements and instrument design can influence the preparation of 3DPT.

5. Conclusions

Within the limitations of the present study, it can be concluded that the SMRP described is effective in removing the support material SUP706B™. PolyJet™ is adequate for printing 3DPT. Furthermore, 3DPT printed with high-temperature RGD525™ have similar behaviour during endodontic preparation with PTG to that of NT, and 3DPT can be used when comparing two preparation systems.

Author Contributions: Conceptualization, T.R., C.B., M.F. and B.M.-B.; methodology, T.R., C.B. and M.F.; software, T.R., M.F. and R.S.; validation, T.R., C.B. and B.M.-B.; resources, T.R., M.F., R.S. and N.A.; data curation, T.R.; writing—original draft preparation, T.R., C.B. and M.F.; writing—review and editing, R.S., N.A., J.M.-C., P.C.-B. and B.M.-B. All authors have read and agreed to the published version of the manuscript.

Funding: This research was funded by Fundação para a Ciência e a Tecnologia and Centro 2020 (UIDB/04044/2020, LA/P/0112/2020).

Institutional Review Board Statement: The study was conducted in accordance with the Declaration of Helsinki and approved by the ethics committee of Fernando Pessoa University (FCS/PI 429/23).

Informed Consent Statement: Informed consent was obtained from all subjects involved in the study.

Data Availability Statement: The data that support the finding of this study are available from the corresponding author upon reasonable request. The data are not publicly available due to privacy and ethical restrictions (undergoing PhD thesis).

Conflicts of Interest: The authors declare no conflicts of interest.

References

1. AAE. Glossary of Endodontic Terms. Available online: https://www.aae.org/specialty/clinical-resources/glossary-endodontic-terms/ (accessed on 4 May 2022).
2. Pinheiro, S.R.; Alcalde, M.P.; Vivacqua-Gomes, N.; Bramante, C.M.; Vivan, R.R.; Duarte, M.A.H.; Vasconcelos, B.C. Evaluation of apical transportation and centring ability of five thermally treated NiTi rotary systems. *Int. Endod. J.* **2018**, *51*, 705–713. [CrossRef]
3. Hulsmann, M. A critical appraisal of research methods and experimental models for studies on root canal preparation. *Int. Endod. J.* **2022**, *55* (Suppl. S1), 95–118. [CrossRef]
4. Reis, T.; Barbosa, C.; Franco, M.; Baptista, C.; Alves, N.; Castelo-Baz, P.; Martin-Cruces, J.; Martin-Biedma, B. 3D-Printed Teeth in Endodontics: Why, How, Problems and Future-A Narrative Review. *Int. J. Environ. Res. Public Health* **2022**, *19*, 7966. [CrossRef]
5. De-Deus, G.; Rodrigues, E.A.; Lee, J.K.; Kim, J.; Silva, E.; Belladonna, F.G.; Simoes-Carvalho, M.; Souza, E.M.; Versiani, M.A. Taper 0.06 Versus Taper 0.04: The Impact on the Danger Zone. *J. Endod.* **2023**, *49*, 536–543. [CrossRef]
6. Arola, D.D.; Gao, S.; Zhang, H.; Masri, R. The Tooth: Its Structure and Properties. *Dent. Clin. N. Am.* **2017**, *61*, 651–668. [CrossRef]
7. Reymus, M.; Fotiadou, C.; Kessler, A.; Heck, K.; Hickel, R.; Diegritz, C. 3D printed replicas for endodontic education. *Int. Endod. J.* **2019**, *52*, 123–130. [CrossRef]
8. Orel, L.; Velea-Barta, O.-A.; Nica, L.-M.; Boscornea-Pușcu, A.-S.; Horhat, R.M.; Talpos-Niculescu, R.-M.; Sinescu, C.; Duma, V.-F.; Vulcanescu, D.-D.; Topala, F.; et al. Evaluation of the Shaping Ability of Three Thermally Treated Nickel–Titanium Endodontic Instruments on Standardized 3D-printed Dental Replicas Using Cone-Beam Computed Tomography. *Medicina* **2021**, *57*, 901. [CrossRef]
9. Kolling, M.; Backhaus, J.; Hofmann, N.; Kess, S.; Krastl, G.; Soliman, S.; Konig, S. Students' perception of three-dimensionally printed teeth in endodontic training. *Eur. J. Dent. Educ.* **2022**, *26*, 653–661. [CrossRef]
10. Karatekin, A.O.; Keles, A.; Gencoglu, N. Comparison of continuous wave and cold lateral condensation filling techniques in 3D printed simulated C-shape canals instrumented with Reciproc Blue or Hyflex EDM. *PLoS ONE* **2019**, *14*, e0224793. [CrossRef]
11. Kooanantkul, C.; Shelton, R.M.; Camilleri, J. Comparison of obturation quality in natural and replica teeth root-filled using different sealers and techniques. *Clin. Oral Investig.* **2023**, *27*, 2407–2417. [CrossRef]
12. Ordinola-Zapata, R.; Bramante, C.M.; Duarte, M.A.; Cavenago, B.C.; Jaramillo, D.; Versiani, M.A. Shaping ability of reciproc and TF adaptive systems in severely curved canals of rapid microCT-based prototyping molar replicas. *J. Appl. Oral Sci.* **2014**, *22*, 509–515. [CrossRef] [PubMed]
13. Lee, K.Y.; Cho, J.W.; Chang, N.Y.; Chae, J.M.; Kang, K.H.; Kim, S.C.; Cho, J.H. Accuracy of three-dimensional printing for manufacturing replica teeth. *Korean J. Orthod.* **2015**, *45*, 217–225. [CrossRef] [PubMed]
14. Moraes, R.D.R.; Santos, T.; Marceliano-Alves, M.F.; Pintor, A.V.B.; Lopes, R.T.; Primo, L.G.; Neves, A.A. Reciprocating instrumentation in a maxillary primary central incisor: A protocol tested in a 3D printed prototype. *Int. J. Paediatr. Dent.* **2019**, *29*, 50–57. [CrossRef] [PubMed]
15. Kent, N.J.; Jolivet, L.; O'Neill, P.; Brabazon, D. An evaluation of components manufactured from a range of materials, fabricated using PolyJet technology. *Adv. Mater. Process. Technol.* **2017**, *3*, 318–329. [CrossRef]
16. Herpel, C.; Tasaka, A.; Higuchi, S.; Finke, D.; Kuhle, R.; Odaka, K.; Rues, S.; Lux, C.J.; Yamashita, S.; Rammelsberg, P.; et al. Accuracy of 3D printing compared with milling—A multi-center analysis of try-in dentures. *J. Dent.* **2021**, *110*, 103681. [CrossRef] [PubMed]
17. Macdonald, N.P.; Cabot, J.M.; Smejkal, P.; Guijt, R.M.; Paull, B.; Breadmore, M.C. Comparing Microfluidic Performance of Three-Dimensional (3D) Printing Platforms. *Anal. Chem.* **2017**, *89*, 3858–3866. [CrossRef] [PubMed]
18. Kessler, A.; Hickel, R.; Reymus, M. 3D Printing in Dentistry-State of the Art. *Oper. Dent.* **2020**, *45*, 30–40. [CrossRef] [PubMed]

19. Stratasys. SUP706 and SUP706B Support Material—EN PolyJet Best Practice. Available online: https://support.stratasys.com/en/Materials/PolyJet/PolyJet-Support (accessed on 27 January 2022).
20. Smutny, M.; Kopecek, M.; Bezrouk, A. An Investigation of the Accuracy and Reproducibility of 3D Printed Transparent Endodontic Blocks. *Acta Medica* **2022**, *65*, 59–65. [CrossRef] [PubMed]
21. Reis, T.; Barbosa, C.; Franco, M.; Batista, C.; Alves, N.; Castelo-Baz, P.; Martin-Cruces, J.; Martin-Biedma, B. Root Canal Preparation of a Commercial Artificial Tooth versus Natural Tooth—A MicroCT Study. *Appl. Sci.* **2023**, *13*, 9400. [CrossRef]
22. Gagliardi, J.; Versiani, M.A.; de Sousa-Neto, M.D.; Plazas-Garzon, A.; Basrani, B. Evaluation of the Shaping Characteristics of ProTaper Gold, ProTaper NEXT, and ProTaper Universal in Curved Canals. *J. Endod.* **2015**, *41*, 1718–1724. [CrossRef]
23. Schneider, S.W. A comparison of canal preparations in straight and curved root canals. *Oral Surg. Oral Med. Oral Pathol.* **1971**, *32*, 271–275. [CrossRef] [PubMed]
24. Silva, E.; Ajuz, N.C.; Martins, J.N.R.; Antunes, B.R.; Lima, C.O.; Vieira, V.T.L.; Braz-Fernandes, F.M.; Versiani, M.A. Multimethod analysis of three rotary instruments produced by electric discharge machining technology. *Int. Endod. J.* **2023**, *56*, 775–785. [CrossRef] [PubMed]
25. Faus-Matoses, V.; Faus-Llacer, V.; Ruiz-Sanchez, C.; Jaramillo-Vasconez, S.; Faus-Matoses, I.; Martin-Biedma, B.; Zubizarreta-Macho, A. Effect of Rotational Speed on the Resistance of NiTi Alloy Endodontic Rotary Files to Cyclic Fatigue-An In Vitro Study. *J. Clin. Med.* **2022**, *11*, 3143. [CrossRef] [PubMed]
26. Perez Morales, M.L.N.; Gonzalez Sanchez, J.A.; Olivieri, J.G.; Elmsmari, F.; Salmon, P.; Jaramillo, D.E.; Terol, F.D. Micro-computed Tomographic Assessment and Comparative Study of the Shaping Ability of 6 Nickel-Titanium Files: An In Vitro Study. *J. Endod.* **2021**, *47*, 812–819. [CrossRef] [PubMed]
27. Serefoglu, B.; Piskin, B. Micro computed tomography evaluation of the Self-adjusting file and ProTaper Universal system on curved mandibular molars. *Dent. Mater. J.* **2017**, *36*, 606–613. [CrossRef] [PubMed]
28. Donnermeyer, D.; Averkorn, C.; Burklein, S.; Schafer, E. Cleaning Efficiency of Different Irrigation Techniques in Simulated Severely Curved Complex Root Canal Systems. *J. Endod.* **2023**, *49*, 1548–1552. [CrossRef] [PubMed]
29. Jiang, L.M.; Verhaagen, B.; Versluis, M.; van der Sluis, L.W. Evaluation of a sonic device designed to activate irrigant in the root canal. *J. Endod.* **2010**, *36*, 143–146. [CrossRef] [PubMed]
30. Haupt, F.; Meinel, M.; Gunawardana, A.; Hulsmann, M. Effectiveness of different activated irrigation techniques on debris and smear layer removal from curved root canals: A SEM evaluation. *Aust. Endod. J.* **2020**, *46*, 40–46. [CrossRef] [PubMed]
31. Provoost, C.; Rocca, G.T.; Thibault, A.; Machtou, P.; Bouillaguet, S. Influence of Needle Design and Irrigant Flow Rate on the Removal of Enterococcus faecalis Biofilms In Vitro. *Dent. J.* **2022**, *10*, 59. [CrossRef]
32. Hartmann, R.C.; Peters, O.A.; de Figueiredo, J.A.P.; Rossi-Fedele, G. Association of manual or engine-driven glide path preparation with canal centring and apical transportation: A systematic review. *Int. Endod. J.* **2018**, *51*, 1239–1252. [CrossRef]
33. Rebong, R.E.; Stewart, K.T.; Utreja, A.; Ghoneima, A.A. Accuracy of three-dimensional dental resin models created by fused deposition modeling, stereolithography, and Polyjet prototype technologies: A comparative study. *Angle Orthod.* **2018**, *88*, 363–369. [CrossRef]
34. Mendřický, R. Accuracy analysis of additive technique for parts manufacturing. *MM Sci. J.* **2016**, *2016*, 1502–1508. [CrossRef]
35. Kim, S.Y.; Shin, Y.S.; Jung, H.D.; Hwang, C.J.; Baik, H.S.; Cha, J.Y. Precision and trueness of dental models manufactured with different 3-dimensional printing techniques. *Am. J. Orthod. Dentofac. Orthop.* **2018**, *153*, 144–153. [CrossRef] [PubMed]
36. Rouze l'Alzit, F.; Cade, R.; Naveau, A.; Babilotte, J.; Meglioli, M.; Catros, S. Accuracy of commercial 3D printers for the fabrication of surgical guides in dental implantology. *J. Dent.* **2022**, *117*, 103909. [CrossRef]
37. Kim, J.H.; Pinhata-Baptista, O.H.; Ayres, A.P.; da Silva, R.L.B.; Lima, J.F.; Urbano, G.S.; No-Cortes, J.; Vasques, M.T.; Cortes, A.R.G. Accuracy Comparison among 3D-Printing Technologies to Produce Dental Models. *Appl. Sci.* **2022**, *12*, 8425. [CrossRef]
38. Bezek, L.; Meisel, N.; Williams, C. Exploring variability of orientation and aging effects in material properties of multi-material jetting parts. *Rapid Prototyp. J.* **2016**, *22*, 826–834. [CrossRef]
39. Chan, C.W.; Romeo, V.R.; Lee, A.; Zhang, C.; Neelakantan, P.; Pedulla, E. Accumulated Hard Tissue Debris and Root Canal Shaping Profiles Following Instrumentation with Gentlefile, One Curve, and Reciproc Blue. *J. Endod.* **2023**, *49*, 1344–1351. [CrossRef]
40. Stratasys, A.K. PolyJet™ Materials Reference Guide. Available online: https://support.stratasys.com/en/materials/polyjet/high-temp (accessed on 8 November 2023).
41. Staninec, M.; Marshall, G.W.; Hilton, J.F.; Pashley, D.H.; Gansky, S.A.; Marshall, S.J.; Kinney, J.H. Ultimate tensile strength of dentin: Evidence for a damage mechanics approach to dentin failure. *J. Biomed. Mater. Res.* **2002**, *63*, 342–345. [CrossRef] [PubMed]
42. Chun, K.; Choi, H.; Lee, J. Comparison of mechanical property and role between enamel and dentin in the human teeth. *J. Dent. Biomech.* **2014**, *5*, 1758736014520809. [CrossRef]
43. Plotino, G.; Grande, N.M.; Bedini, R.; Pameijer, C.H.; Somma, F. Flexural properties of endodontic posts and human root dentin. *Dent. Mater.* **2007**, *23*, 1129–1135. [CrossRef]
44. Aazzouzi-Raiss, K.; Ramirez-Munoz, A.; Mendez, S.P.; Vieira, G.C.S.; Aranguren, J.; Perez, A.R. Effects of Conservative Access and Apical Enlargement on Shaping and Dentin Preservation with Traditional and Modern Instruments: A Micro-computed Tomographic Study. *J. Endod.* **2023**, *49*, 430–437. [CrossRef]

45. Duque, J.A.; Vivan, R.R.; Cavenago, B.C.; Amoroso-Silva, P.A.; Bernardes, R.A.; Vasconcelos, B.C.; Duarte, M.A. Influence of NiTi alloy on the root canal shaping capabilities of the ProTaper Universal and ProTaper Gold rotary instrument systems. *J. Appl. Oral Sci.* **2017**, *25*, 27–33. [CrossRef]
46. Sivakumar, M.; Nawal, R.R.; Talwar, S.; Baveja, C.P.; Kumar, R.; Yadav, S.; Kumar, S.S. Shaping, and disinfecting abilities of ProTaper Universal, ProTaper Gold, and Twisted Files: A correlative microcomputed tomographic and bacteriologic analysis. *Endodontology* **2023**, *35*, 54–59.
47. Silva, E.; Martins, J.N.R.; Lima, C.O.; Vieira, V.T.L.; Braz Fernandes, F.M.; De-Deus, G.; Versiani, M.A. Mechanical Tests, Metallurgical Characterization, and Shaping Ability of Nickel-Titanium Rotary Instruments: A Multimethod Research. *J. Endod.* **2020**, *46*, 1485–1494. [CrossRef] [PubMed]
48. Yalniz, H.; Koohnavard, M.; Oncu, A.; Celikten, B.; Orhan, A.I.; Orhan, K. Comparative evaluation of dentin volume removal and centralization of the root canal after shaping with the ProTaper Universal, ProTaper Gold, and One-Curve instruments using micro-CT. *J. Dent. Res. Dent. Clin. Dent. Prospect.* **2021**, *15*, 47–52. [CrossRef] [PubMed]
49. Peters, O.A.; Laib, A.; Gohring, T.N.; Barbakow, F. Changes in root canal geometry after preparation assessed by high-resolution computed tomography. *J. Endod.* **2001**, *27*, 1–6. [CrossRef] [PubMed]
50. Romeiro, K.; Brasil, S.C.; Souza, T.M.; Gominho, L.F.; Perez, A.R.; Perez, R.; Alves, F.R.F.; Rocas, I.N.; Siqueira, J.F., Jr. Influence of brushing motions on the shaping of oval canals by rotary and reciprocating instruments. *Clin. Oral Investig.* **2023**, *27*, 3973–3981. [CrossRef] [PubMed]
51. Hilaly Eid, G.E.; Wanees Amin, S.A. Changes in diameter, cross-sectional area, and extent of canal-wall touching on using 3 instrumentation techniques in long-oval canals. *Oral Surg. Oral Med. Oral Pathol. Oral Radiol. Endod.* **2011**, *112*, 688–695. [CrossRef] [PubMed]
52. Ha, J.H.; Park, S.S. Influence of glide path on the screw-in effect and torque of nickel-titanium rotary files in simulated resin root canals. *Restor. Dent. Endod.* **2012**, *37*, 215–219. [CrossRef]
53. Ha, J.H.; Kwak, S.W.; Sigurdsson, A.; Chang, S.W.; Kim, S.K.; Kim, H.C. Stress Generation during Pecking Motion of Rotary Nickel-titanium Instruments with Different Pecking Depth. *J. Endod.* **2017**, *43*, 1688–1691. [CrossRef]
54. Kyaw, M.S.; Ebihara, A.; Kasuga, Y.; Maki, K.; Kimura, S.; Htun, P.H.; Nakatsukasa, T.; Okiji, T. Influence of rotational speed on torque/force generation and shaping ability during root canal instrumentation of extracted teeth with continuous rotation and optimum torque reverse motion. *Int. Endod. J.* **2021**, *54*, 1614–1622. [CrossRef] [PubMed]
55. Kimura, S.; Ebihara, A.; Maki, K.; Nishijo, M.; Tokita, D.; Okiji, T. Effect of Optimum Torque Reverse Motion on Torque and Force Generation during Root Canal Instrumentation with Crown-down and Single-length Techniques. *J. Endod.* **2020**, *46*, 232–237. [CrossRef] [PubMed]

Disclaimer/Publisher's Note: The statements, opinions and data contained in all publications are solely those of the individual author(s) and contributor(s) and not of MDPI and/or the editor(s). MDPI and/or the editor(s) disclaim responsibility for any injury to people or property resulting from any ideas, methods, instructions or products referred to in the content.

Article

Influence of Cold Atmospheric Plasma on Surface Characteristics and Bond Strength of a Resin Nanoceramic

Xiaoming Zhu [1,2,†], Jiamin Shi [3,4,†], Xinyi Ye [2,3], Xinrong Ma [5], Miao Zheng [5], Yang Yang [2,3,*] and Jianguo Tan [2,3,*]

1. Second Clinical Division, Peking University School and Hospital of Stomatology, Beijing 100101, China
2. National Center of Stomatology, National Clinical Research Center for Oral Diseases, National Engineering Laboratory for Digital and Material Technology of Stomatology, Beijing Key Laboratory of Digital Stomatology, Beijing 100081, China
3. Department of Prosthodontics, Peking University School and Hospital of Stomatology, Beijing 100081, China
4. Guanghua School of Stomatology, Sun Yat-Sen University, Guangzhou 510055, China
5. Department of Stomotology, Peking University Third Hospital, Beijing 100191, China
* Correspondence: yyangpkuss@163.com (Y.Y.); tanwume@bjmu.edu.cn (J.T.)
† These authors contributed equally to this work.

Citation: Zhu, X.; Shi, J.; Ye, X.; Ma, X.; Zheng, M.; Yang, Y.; Tan, J. Influence of Cold Atmospheric Plasma on Surface Characteristics and Bond Strength of a Resin Nanoceramic. *Materials* 2023, *16*, 44. https://doi.org/10.3390/ma16010044

Academic Editors: Luigi Generali, Vittorio Checchi, Eugenio Pedullà and Gianrico Spagnuolo

Received: 8 November 2022
Revised: 7 December 2022
Accepted: 16 December 2022
Published: 21 December 2022

Copyright: © 2022 by the authors. Licensee MDPI, Basel, Switzerland. This article is an open access article distributed under the terms and conditions of the Creative Commons Attribution (CC BY) license (https://creativecommons.org/licenses/by/4.0/).

Abstract: The purpose of this study was to investigate the effect of cold atmospheric plasma (CAP) treatment on resin nanoceramic (RNC) surface state and its bond strength with resin cement. RNC with different surface treatments were prepared: control, sandblasting treatment (SB), hydrofluoric acid etching (HF) and plasma treatment of helium gas (CAP-He) and argon gas (CAP-Ar). The prepared samples were measured by SEM, Ra, Rz, contact angle goniometer, and XPS for surface characteristics. The shear bond test of RNC was examined in nine groups: SB + saline coupling agent (SL), HF + SL, CAP-He/Ar, CAP-He/Ar + SL, SB + CAP-He/Ar + SL, and control. The bond strength between RNC and resin cement was compared using shear bond strength test, before and after thermocycling. After CAP irradiation, the surface topography maintained, while the surface water contact angle was significantly reduced to $10.18° \pm 1.36°$ (CAP-He) and $7.58° \pm 1.79°$ (CAP-Ar). The removal of carbon contamination and inducing of oxygen radicals was detected after CAP treatment. The bond strength was improved by CAP treatment, but varied on CAP gas species and combination methods. CAP of Ar gas had better SBS than He gas. After thermocycling, CAP-Ar + SL showed the maximized shear bond strength (32.38 ± 1.42 MPa), even higher than SB + SL group (30.08 ± 2.80 MPa, $p < 0.05$). In conclusion, CAP treatment of helium and argon can improve the bonding properties of RNC by improving surface wettability, and CAP of argon gas combined with silane coupling agent shows the highest bond strength.

Keywords: resin nanoceramic; shear bond strength; cold atmospheric plasma; surface wettability

1. Introduction

Resin-matrix ceramic is a new class of dental esthetic restorative material, consisting of a resin matrix filled with ceramic particles [1,2]. It has similar compositions and mechanical properties of both resin and ceramic, as "ceramic-like materials" [3,4]. In recent years, CAD/CAM resin-matrix ceramic materials have been developed rapidly, and widely used in indirect prosthesis, such as inlays, onlays and crowns, with easier and faster manufacturing [5,6]. However, for the long-term success of CAD/CAM restorations, great bonding between the restorative material and resin cement are required. Concerns have been raised regarding debonding and fracture rate of resin-matrix ceramic restorations, which may weaken their clinical performance [7,8].

According to different composition and microstructure, CAD/CAM resin-matrix ceramic materials are mainly divided into two types: polymer-infiltrated hybrid ceramic (PIHC) and resin nanoceramic (RNC) [2,3]. For PIHC, it's typically composed of a dual network: a feldspathic ceramic network and a poly network [3]. It has been demonstrated great bonding properties after etching with hydrofluoric acid, similar as feldspathic ceramics [8,9]. While for RNC, it consists of a highly cured resin matrix reinforced with silica or zirconia nanoparticles. Since RNC is non-ecthable material, sandblasting combined with silane coupling agent has been recommended before its bonding [10–12]. However, sandblasting may damage the structure of resin matrix, decreasing its flexural strength, even causing fracture [13]. Is there an effective surface treatment method for improving the bonding performance of RNC without damaging its mechanical strength?

Cold atmospheric plasma (CAP), as an efficient and clean surface treatment method, has been used to improve the bonding properties of other all-ceramic dental materials, like zirconia and glass-matrix ceramics [14–17]. It can effectively modify the physicochemical properties of those materials, by increasing their hydrophilicity [18,19]. The bond strength of RNC can also be enhanced by increasing its surface energy, wettability and reactivity [20,21]. Whether CAP treatment an effective method of RNC bonding? More supported evidence is still needed.

The purpose of this study was to evaluate the effect and mechanism of CAP on resin bonding to RNC, comparing with sandblasting and hydrofluoric acid etching. Physical and chemical alterations of RNC after treatment were assessed, as well as the shear bond strength (SBS) before and after thermocycling. The first null hypothesis was that CAP treatment did not change the surface characteristics of RNC and improve the bond strength of RNC. Besides, CAP of different kinds of gas, like helium or argon, and combinations with sandblasting were also evaluated in this study. The second null hypothesis was that there's no difference between various CAP methods. This study was expected to prove CAP as a new method improving RNC's bonding properties. Besides, the mechanism of CAP demonstrated in this study may broaden its application in dental materials' modification.

2. Materials and Methods

2.1. Preparation of RNC Specimens

The RNC used in this study was Renci CAD/CAM resin nano ceramic (UPCERA, Shenzhen, China) widely used in China in recent years [1]. The main composition of RNC and other materials used in this study was listed in Table 1. The RNC blocks were sectioned into square specimens (14 mm × 12 mm × 2 mm), using a slow-speed diamond wafering blade (Isomet 1000 Precision Saw, Buehler; Lake Bluff, IL, USA). They were then wet ground (Automet 500; Buehler, Esslingen, Germany) by 600-grit SiC for about 1 min. Before further experiments, all specimens were ultrasonically cleaned and thoroughly dried. The schematic design of the study was shown in Figure 1.

Table 1. Materials used in this study with their manufactures and main compositions.

Materials	Manufacture	Main Composition *
resin nanoceramic	Renci Upcera	13–43 wt% polymer, 55–85 wt% glass ceramic (including 0.1 μm~1 μm B_2O_3, 5 nm~50 nm SiO_2)
hydrofluoric acid	IPS ceramic etching gel Ivoclar Vivadent	≤ 5% hydrofluoric acid
silane coupling agent	RelyX ceramic primer 3M ESPE	Ethanol, 3-trimethoxysilylpropyl methacrylate
composite resin	Clearfil AP-X Kuraray	Base resin: Bis-GMA, TEGDMA Filler: 85 wt% silanated barium glass filler of irregular shape (700 nm), and silanated silica filler (100–1500 nm)
resin cement	RelyX U200 3M ESPE	Bi-functional (meth) acrylate; Inorganic fillers (43% by volume)

Bis-GMA: Bisphenol A glycerolate dimethacrylate; TEGDMA: Triethylene glycol dimethacrylate;
* The information was provided by the manufacturers.

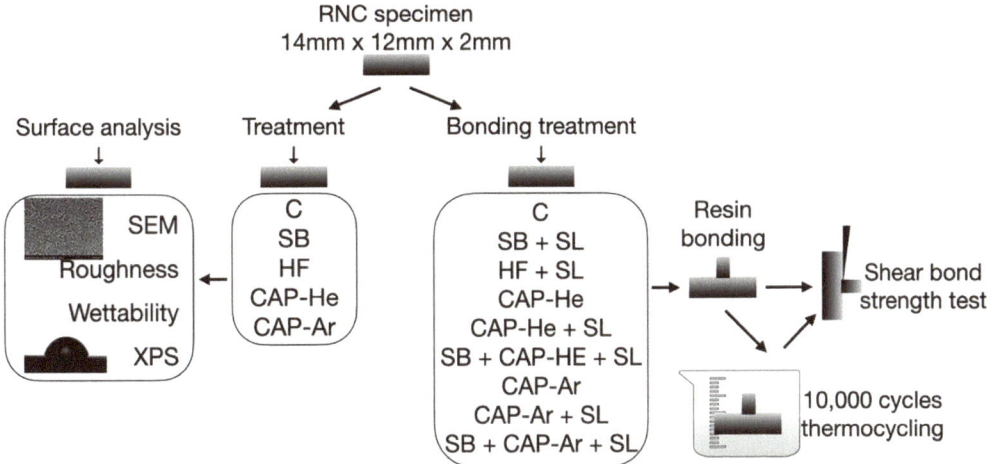

Figure 1. The schematic design of the study.

2.2. Surface Treatment and Analysis

All RNC specimens were submitted into five groups for treatment:

- Control group (C): specimen with no treatment.
- Sandblasting group (SB): 50 μm Al_2O_3 particles (COBRA, Rengert, Germany) were sandblasted for 20 s at a pressure of 0.1 MPa and a distance of 10 mm. After sand blasting, ultrasonically cleaning (using deionized water for 5 min) and drying was carried out for the specimens.
- Hydrofluoric acid etching group (HF): A hydrofluoric acid agent (IPS Ceramic Etching Gel, Ivoclar Vivadent, Schaan, Liechtenstein) was applied for 60 s with a disposable brush, rinsed with deionized water for 1 min and then thoroughly dried.
- CAP jet with helium gas group (CAP-He): specimen was treated with a CAP jet with a helium flow rate for 120 s, at a distance of 10 mm.
- CAP jet with argon gas group (CAP-Ar): The surface was treated with a CAP jet with an argon flow rate for 120 s, at a distance of 10 mm.

In both CAP groups, the plasma was produced by CAP Med-I (Figure 2) at the condition of 2.8 kV and 17 kHz, with the gas flow rate of 8.1 slpm. The other details of this equipment have been reported before [22–24].

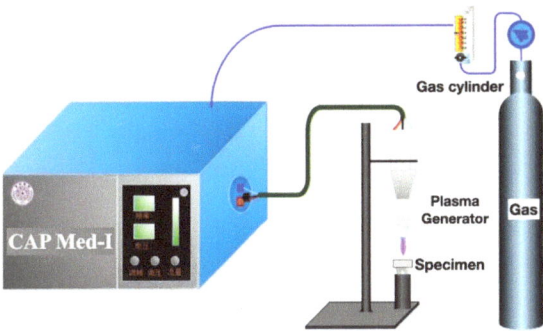

Figure 2. CAP-Med I equipment used in this study for plasma jet generation.

After treatment, specimens in all five groups (five specimens per group) were first observed in a scanning electron microscopy (SEM) (S-4800, Hitachi, Japan) at 35 mA for 85 s, after sputter-coated with Au-Pd alloy. The surface roughness µm) of specimen was measured by a stylus surface profilometer (SJ-401, Mitutoyo, Japan), including Ra (arithmetical mean height, in µm) and Rz (maximum height of surface roughness profile, in µm). They were both determined with a cut-off value of 0.8 mm, measurement length of 4 mm. For each specimen, three times was measured at different areas and five specimens per group.

The wettability of surface to water was determined by a contact angle goniometer (SL200, USA Kino Industry, Norcross, GA, USA). Static contact angle was measured using tangential line method. On each specimen, 1 µL deionized water droplet was applied with an automatic piston syringe. Three randomly selected points on each specimen were examined, and five specimens per group.

Electron spectroscopy for chemical analysis (ESCA) was used to evaluate the composition of the treated surface. It was conducted by X-ray photoelectron spectroscopy (Kratos Analytical, Manchester, UK) to evaluate the intensity of carbon, oxygen and silicon on the surface. The binding energy of each spectrum was calibrated with C1s (284.8 eV).

2.3. Shear Bond Strength Test

180 prepared RNC specimens were randomly divided into nine groups according to different surface treatments (Table 2).

Table 2. Different surface treatment of nine groups before binding.

Groups	Surface Treatment before Bonding
control	no treatment
SB + SL	sandblasting + silane coupling agent *
HF + SL	hydrofluoric acid + silane coupling agent
CAP-He	CAP jet with helium gas
CAP-He + SL	CAP jet with helium gas + silane coupling agent
SB + CAP-He + SL	sandblasting + CAP jet with helium gas + silane coupling agent
CAP-Ar	CAP jet with argon gas
CAP-Ar + SL	CAP jet with argon gas + silane coupling agent
SB + CAP-Ar + SL	sandblasting + CAP jet with argon gas + silane coupling agent

* RelyX ceramic primer (3M ESPE, St. Paul, MN, USA) as silane coupling agent was applied to RNC specimens for 1 min, and dried with oil-free air spray.

A light-cured resin material (Clearfil AP-X, Kuraray, Tokyo, Japan) was fabricated intio resin cylinders (4 mm in diameter and 4 mm in height). The process has been reported in our previous study [17]. After prepared, the resin cylinder was bonded onto group-treated RNC's surface, using self-adhesion resin cement (RelyX U200 3M ESPE, St. Paul, MN, USA) at a static load of 5 N. 40 s of LED light irradiation was applied, before cement excess was carefully removed.

After bonding, half of samples in each group were submitted to the shear bond strength test immediately (n = 10), and the other half were submitted to a thermocycling aging for 10,000 cycles (5 °C~55 °C), before shear bond strength test (n = 10).

The shear fracture loads were measure by a universal mechanical testing machine (EZ-L, SHIMADZU, Japan). The crosshead had a speed of 1 mm/min until failure. The SBS was calculated as:

$$\text{SBS (MPa)} = \text{Maximum load}, F\text{ (N)}/ \text{ Bonding Area}, S \text{ (mm}^2)$$

The failure mode after shearing was observed (SMZ-10, Nikon, Japan) and classified as: adhesive failures, mixed failures and cohesive failures.

2.4. Statistical Analysis

All the data were collected and expressed as means and standard deviations, after normal distribution test. They were tested for statistical significance using one-way ANOVA analysis of variance, with a significance level of 0.05. Post-hoc analysis using the Tukey method was performed to detect pairs of groups with statistically significant differences. The data were statistically analyzed using SPSS software version 25.0 (SPSS, IBM Corp., Chicago, IL, USA).

3. Results
3.1. Surface Analysis of RNC after Treatment

The surface morphology of RNC in different groups was shown in Figure 3. Compared to control group, SB presented a roughed surface with several grooves and faceted slits, embedded with few particles. While HF produced a slightly porous morphology, with glassy matrix removal. Both CAP-He and CAP-Ar did not result in visible change to RNC surface.

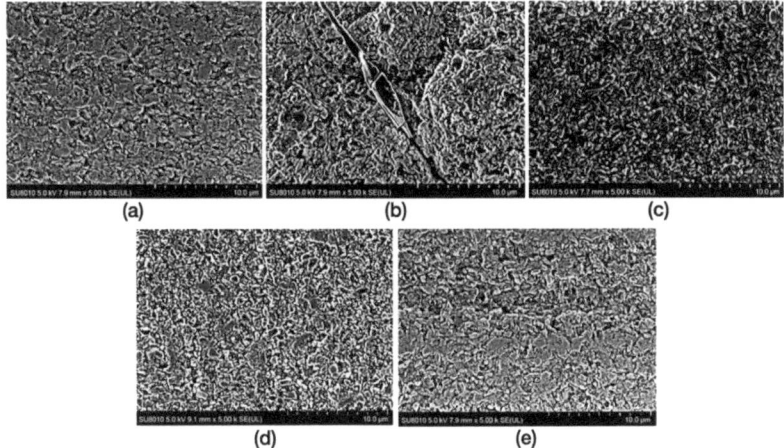

Figure 3. Scanning electron microscopy images of RNC specimens. (**a**) specimen in control group; (**b**) specimen in SB treatment; (**c**) specimen in HF treatment; (**d**) specimen in CAP-He treatment; (**e**) specimen in CAP-Ar treatment.

Surface roughness of RNC specimen was presented in Table 3. The Ra value of untreated RNC was 0.09 ± 0.01 μm. There's no significant difference after CAP-He and CAP-Ar treated (0.09 ± 0.02 μm for CAP-He, and 0.10 ± 0.02 μm for CAP-Ar, $p > 0.05$). After SB and HF treatment, both Ra values increased (0.47 ± 0.05 μm and 0.17 ± 0.02 μm respectively, $p < 0.05$). The Rz value of CAP-Ar (0.79 ± 0.14 μm) was slightly larger than CAP-He (0.75 ± 0.13 μm), but with no significant difference ($p = 0.99$). The Rz values of other groups showed the same trend.

Table 3. Surface roughness of RNC specimens.

Groups	Ra (Mean ± SD, μm)	Rz (Mean ± SD, μm)
control	0.09 ± 0.01 [c]	0.71 ± 0.12 [C]
SB	0.47 ± 0.05 [a]	3.87 ± 0.33 [A]
HF	0.17 ± 0.02 [b]	1.38 ± 0.13 [B]
CAP-He	0.09 ± 0.02 [c]	0.75 ± 0.13 [C]
CAP-Ar	0.10 ± 0.02 [c]	0.79 ± 0.14 [C]

Ra: arithmetical mean height, Rz: maximum height of surface roughness profile. Different letters in columns showed significant differences ($p < 0.05$) between groups.

The water contact angle measurements were presented in Figure 4. Compared to the high value of un-treated RNC ($70.05° \pm 0.86°$), HF decreased the water contact angle into $58.64° \pm 2.72°$ ($p < 0.05$), while SB increased the value ($96.21° \pm 2.03°$, $p < 0.05$). While, for both CAP groups, the water contact angles decreased significantly ($p < 0.01$), $10.18° \pm 1.36°$ for CAP-He, and $7.58° \pm 1.79°$ for CAP-Ar, showing the greatly enhanced hydrophilicity.

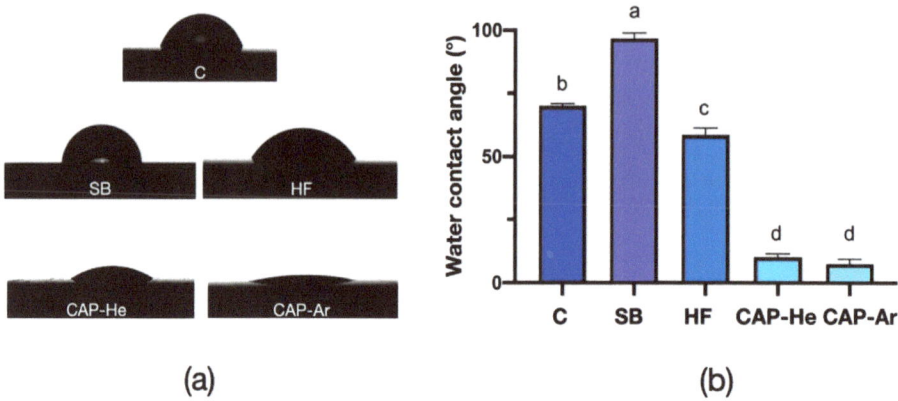

Figure 4. Water contact angle of RNC specimens. (a) classical images of each group; (b) statistical analysis of water contact angle values, and different letters denote significant differences ($p < 0.05$).

The XPS analysis of RNC specimens showed peaks of C 1s, O 1s, N 1s, Ba 3d5, Si 2p (Figure 5). Table 4 showed the main chemical compositions (C 1s, O 1s, Si 2p) and C/O ratio of each group. After the CAP-He and CAP-Ar treatment, the content of C1s decreased significantly from 51.17% to 29.12% and 28.14%, while the oxygen content increased from 36.90% to 50.93% and 51.2%. C/O ratio was 0.57 for CAP-He, and 0.55 for CAP-Ar. Besides, the silicon content decreased after hydrofluoric acid.

Table 4. Main chemical compositions on RNC surfaces.

Groups	Atomic %			C/O
	C 1s	O 1s	Si 2p	
C	51.17	36.90	9.75	1.39
SB	56.30	33.94	7.69	1.66
HF	69.68	27.50	0.09	2.53
CAP-He	29.12	50.93	16.62	0.57
CAP-Ar	28.14	51.2	17.82	0.55

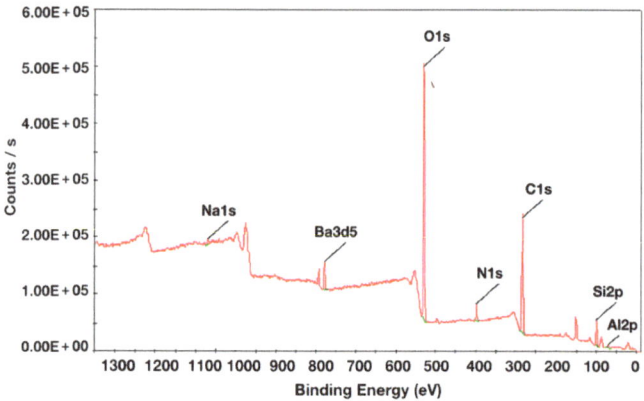

Figure 5. XPS wide-scan spectra of RNC specimen.

3.2. Shear Bond Strength of RNC

As listed in Table 5, the immediate shear bond strengths of RNC in all treatment groups were significantly higher than control group (16.83 ± 1.77 MPa, $p < 0.05$). Groups SB + SL, SB + CAP-He + SL, CAP-Ar, CAP-Ar + SL and SB + CAP-Ar + SL all presented the comparably SBS, presenting the highest values among all treated groups ($p < 0.05$). HF + SL presented the lowest SBSs (21.17 ± 1.37, $p < 0.05$) among all treatment groups.

Table 5. Shear bond strength (SBS) of RNC specimens in different groups.

Groups	SBSs (Mean ± SD, MPa)	
	Immediate (n = 10)	After Thermocycling (n = 10)
control	16.83 ± 1.77 [d]	13.38 ± 3.90 [F*]
SB + SL	31.50 ± 2.74 [a]	30.08 ± 2.80 [B]
HF + SL	21.17 ± 1.37 [c]	17.47 ± 2.04 [E*]
CAP-He	27.93 ± 1.74 [b]	23.02 ± 2.62 [D*]
CAP-He + SL	28.51 ± 1.71 [b]	25.98 ± 1.74 [C*]
SB + CAP-He + SL	32.10 ± 2.57 [a]	28.77 ± 3.07 [B,C*]
CAP-Ar	33.97 ± 2.04 [a]	28.28 ± 2.56 [B,C*]
CAP-Ar + SL	33.78 ± 1.60 [a]	32.38 ± 1.42 [A]
SB + CAP-Ar + SL	32.52 ± 2.31 [a]	28.90 ± 1.82 [B,C*]

Different letters in columns showed significant differences ($p < 0.05$) between groups. * Represented a significant difference ($p < 0.05$) between immediate and after thermocycling SBSs.

After 10,000 thermocycling, the binding area of samples maintained the same, while SBS values of treated groups were all significantly higher than control (13.38 ± 3.90 MPa, $p < 0.05$). Group CAP-Ar + SL (32.38 ± 1.42 MPa) exhibited the highest SBSs ($p < 0.05$) among all treatment groups, while groups SB + SL, SB + CAP-He + SL, CAP-Ar and SB + CAP-Ar + SL presented the comparable SBS ($p > 0.05$). Compared with immediate shear bond strength, the SBS values in most groups decreased significantly after thermocycling ($p < 0.05$). While only in SB + SL and CAP-Ar + SL, SBSs after aging maintained the comparable values ($p > 0.05$) compared with their immediate SBSs.

Due to the mechanical properties of RNC, the failure modes were classified as: adhesive failure (fractured at the RNC/resin cement bonding interface), mixed failure (fractured occurred at both bonding interface and inside resin cement) and cohesive failure (fractured inside RNC specimens or resin cement). The failure modes (Figure 6) in control group were all adhesive failures. In SB + SL group, before and after thermocycling, the failure mode showed large proportion of cohesive fractures predominantly within the RNC specimens. There're also cohesive fractures in groups SB + CAP-He + SL and SB + CAP-Ar + SL. In other treated groups, there were only mixed and adhesive failures, and the percentage of adhesive failures increased after thermocycling.

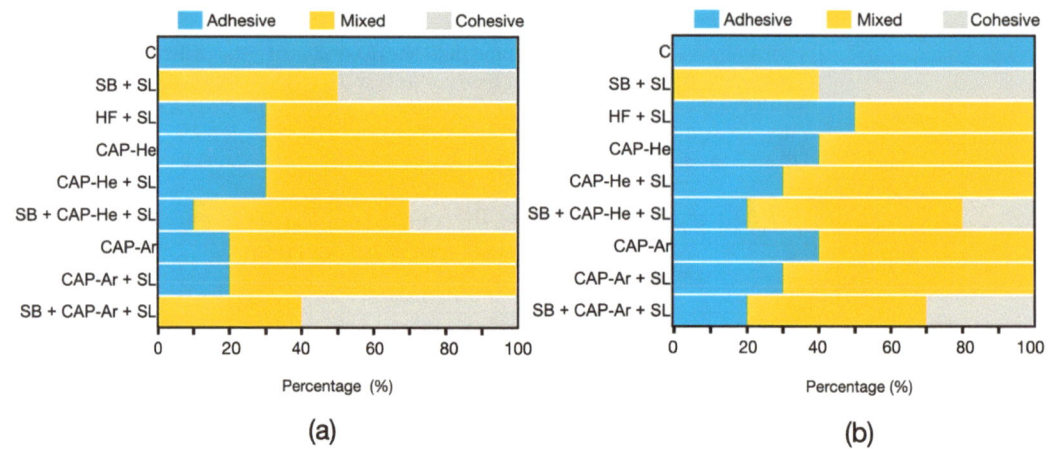

Figure 6. Failure mode distributions of shear bond test before (**a**) and after (**b**) thermocycling.

4. Discussion

Great and endurable bonding with resin cement is prominent for dental hybrid and ceramic materials. As a novel CAD/CAM material, RNC has displayed great clinical performance with relatively high bonding strength [2–6]. The present study evaluated the effect of helium and argon CAP treatment on surface characteristics and bond strength of RNC. According to the results, the first null hypothesis stating that CAP treatment dos not change RNC's surface characteristics and bond strength should be rejected. Moreover, the second null hypothesis of the study stating that there's no difference between helium or argon CAP or combinations of CAP treatment should also be rejected.

For RNC, consisting of a highly cured resin matrix reinforced with silica nanoparticles, controlled sandblasting and silane coupling agent was recommended before bonding rather than acid [11,12,25]. After sandblasting, the roughened surface may allow resin cement to flow into these micro-retentions, and form a stronger micro-mechanical interlock. However, it has been reported that sandblasting could initiate surface defects that may compromise the mechanical properties of resin nanoceramic materials [13]. Tekçe et al. [25] also demonstrated that excessive sandblasting (50 μm alumina 0.2 MPa for 30 s) produced large crack propagating along the material, and also decreased the microtensile test values after 5000 thermocycling. The residual sandblasting particles may also affect its bonding properties. Ultrasonic cleaning was necessary after sandblasting for unless 5 min [10]. In this present study, the mild sandblasting (50 μm alumina at 0.1 MPa for 20 s) was used, according to the references [10–13,25] and our pre-test. The SEM images presented a roughed surface with several grooves and faceted slits embedded with few particles. Combining with silane coupling agent, it improved the shear bond strength more than hydrofluoric acid, and there's no significant difference between immediate and after thermocycling. But the failure modes of specimens in SB + SL group exhibited a large proportion of cohesive failures. Cohesive fractures predominantly within the RNC revealed that the shear resistance of RNC itself, weakened by sandblasting, was lower than the bonding strength with resin cement. Nevertheless, group SB + SL, with mild sandblasting, was still designed as the gold standard in our shear bond strength test to compare the other pre-bonding methods.

Hydrofluoric acid etching, by dissolving the glassy matrix containing silica (SiO_2), also increases the surface roughness of ceramic materials [9]. It can improve the micromechanical retention of the applied primer, and also enhance surface's wettability, promoting an optimal bond strength. According to the results, after HF acid etching for 60 s, RNC surface became porous and XPS presented significantly decreased Si content from 9.75% to 0.09%. However, in shear bond test, HF + SL led to a relatively low SBS value. This may be related to the chemical structure of RNC, less feldspar ceramic phase and much more polymeric phase, which is less susceptible to hydrofluoric acid than PHIC [3].

CAP is an artificial plasma created by partially ionized gas. It can increase surface energy and hydrophilicity, by generating highly reactive particles, such as ions, electrons and free radicals. It has been demonstrated an effective way to modify surface of dental ceramics and natural tooth for better hydrophilicity [23,26,27]. In this present study, CAP of He and Ar gas both enhanced RNC's wettability significantly. CAP-Ar even modified RNC surface into super hydrophilicity (<10°), while the surface morphology maintained the same. The XPS analysis showed a significant decrease in the C% after He and Ar CAP treatment. The same phenomenon was reported by Henningsen et al. [28] that plasma was capable of decreasing carbon-rich contaminants such as C-OH, C = O, COOH radicals, which are known to compromise surface's bonding strength. Furthermore, the CAP treatment on RNC surface produced an increase in the oxygen content. A high level of reactive -O radicals (i.e., hydroxyl free radicals, excited oxygen ions, and atomic oxygen) can be produced by plasma [17,22,29], thereby highly increasing materials' hydrophilicity. The C/O ratio, representing surface wettability, was 0.57 for CAP-He and 0.55 for CAP-Ar, both much lower than control and other treated RNCs. It demonstrated that the collisions between RNC surface and reactive oxygen species, combining with the reduced carbon content, both contributed to the enhancing of hydrophilicity.

According to the hydrophobic recovery theory [30], when a plasma-treated surface is exposed to the atmosphere, the reactive specimens will react with elements and impurities in the atmosphere, and diminish over time [22,29]. Barquete et al. reported that the improved bonding properties was available only if the surface was cemented within 8 h after CAP irradiation [29]. So, in this study, the bonding process was immediately followed the CAP treatment. Since sandblasting was the gold standard and suggested by manufacture, CAP treatment was designed in this study as a substitute for sandblasting. This present study was mainly focused on whether CAP was a comparable or even better method than sandblasting. The application of CAP has three modes for each gas: CAP alone, CAP combined with silane coupling agent, and CAP combined with SB + SL (CAP treatment after sandblasting, and then silane coupling agent).

Although the wettability of the RNC surface was maximized when plasma irradiation was performed using both helium and argon gas, the SBS values of CAP-He and CAP-Ar were not both maximized. After CAP-He treatment, the bonding strength of RNC has been improved a little, only SB + CAP-He + SL achieved the equal SBS value to SB + SL. While after thermocycling, all those application modes had lower SBSs than gold standard (SB + SL). Fortunately, CAP-Ar had better results. All groups with CAP-Ar had comparably maximized SBSs. Besides, after thermocycling, CAP + SL produced the highest SBS value among all groups. This positive result of may be related to the micro-etching effect and a large amount of oxygen particles, especially -OH radicals, produced by CAP of argon [31–33]. During the discharge with high intensity, water in the air may be ionized into -OH radicals, promoting super-hydrophilic surface [33]. Furthermore, argon plasma can also promote polymerization initiation of resin-based materials [34], promoting better properties and bonding performance of dental composite restorations. To illustrate the different mechanism of argon and helium CAP on RNC, further investigation like Atomic Force Microscope (AFM) or Raman spectra may be needed. Different reactive particles or radicals should be further examined. Nevertheless, it has been demonstrated that micromechanical roughening could increase CAD/CAM hybrid materials' bond strength, more than other chemical modifications, especially after thermocycling aging [35,36]. The study of Castro EF et al. [37] also presented a negative result that no significant benefit was found in RNC's bonding, using plasma alone or combined with a bonding agent. The possible reason might be different CAP devices and conditions. In their study, plasma was generated through a hand-held unit at a flow rate of 1 slpm, and was applied for only 30 s. While in this present study, the intensity of plasma was much greater with a flow rate of 8.1 slpm, and was applied for 120 s. Besides, the RNC used in this study contains

55–85 wt% glass ceramic, including 0.1 μm~1 μm B_2O_3 and 5 nm~50 nm SiO_2. It has been widely used in China. While there's another widely-used RNC, the Lava Ultimate (3M ESPE, Seefeld, Germany), which contains SiO_2, ZrO_2 and aggregated ZrO_2/SiO_2 cluster. Different ceramic filler particles in these RNC materials may also lead to differential reaction to CAP treatment. Besides, Ahn JJ et al. reported a greater bond strength of zirconia after CAP treatment combined with sandblasting [38]. While in this study, for RNC, CAP treatment combined with sandblasting did not promote better bond strength. It may be related to the high content of crystalline in zirconia. The air-abrasion can produce a roughed surface, providing more opportunities for CAP reaction and a larger bonding area, without weakening zirconia's structure. But for RNC, as already said, sandblasting could initiate damage of structure that may compromise the mechanical strength and bonding properties [13,25]. In this study, after thermocycling, CAP-Ar combined with SB + SL had lower SBS values and more cohesive failures than CAP-AR + SL.

Although the present study showed some interesting and meaningful aspects regarding the influence of helium and argon CAP treatment on the bonding performance of RNC. The limitations should be noted that only one kind of RNC materials was used in this study, and only self-adhesive resin cement was tested. The other resin cements, multi-functional primer agent may also affect bonding properties of RNC. Besides, the thermocycling in vitro cannot imitate the real aging performance in clinical situations. Although it has been reported that approximately 10,000 thermo cycles related to 1 year life in vivo [39]. The simulation of mastication forces and saliva both can influence the long-term success of bond, which should be addressed in future investigations.

5. Conclusions

The effectiveness of cold atmospheric plasma irradiation for RNC surface treatment was demonstrated. Plasma treatment with helium and argon gas do not change the surface morphology of RNC, but can significantly improve the surface wettability of this material, by removing carbon contamination and introducing active oxygen radicals. Different CAP treatments resulted in differential bonding properties, and the argon CAP combined with silane coupling agent improved the highest bonding strength. It was suggested that CAP could be a new surface treatment method for RNC bonding.

Author Contributions: Conceptualization, Y.Y. and J.T.; methodology, X.M. and M.Z.; formal analysis, J.S. and Y.Y.; investigation, X.Z. and X.Y.; resources, X.Y. and M.Z.; data curation, X.Z., J.S., X.Y. and X.M.; writing—original draft preparation, X.Z. and J.S.; writing—review and editing, Y.Y. and J.T.; visualization, X.M.; supervision, J.T.; project administration, J.T.; funding acquisition, X.Z., Y.Y. and J.T. All authors have read and agreed to the published version of the manuscript.

Funding: Study supported in part by the National Natural Science Foundation of China (No. 81901033, 82001100 and 82210018) and Beijing Municipal Natural Science Foundation (No. 7212138).

Institutional Review Board Statement: Not applicable.

Informed Consent Statement: Not applicable.

Data Availability Statement: Not applicable.

Acknowledgments: The authors are grateful to Heping Li and his team from Tsinghua University, Department of Engineering Physics, for technical support with CAP equipment and treatment.

Conflicts of Interest: The authors have no conflict of interest relevant to this article.

References

1. Society of Esthetic Dentistry; Chinese Stomatological Association. Expert consensus on clinical application of all-ceramic esthetic restorative materials. *Zhonghua Kou Qiang Yi Xue Za Zhi* **2019**, *54*, 825–828.
2. Fathy, H.; Hamama, H.H.; El-Wassefy, N.; Mahmoud, S.H. Clinical performance of resin-matrix ceramic partial coverage restorations: A systematic review. *Clin. Oral Investig.* **2022**, *26*, 3807–3822. [CrossRef] [PubMed]
3. Gracis, S.; Thompson, V.P.; Ferencz, J.L.; Silva, N.R.; Bonfante, E.A. A new classification system for all-ceramic and ceramic-like restorative materials. *Int. J. Prosthodont.* **2015**, *28*, 227–235. [CrossRef] [PubMed]

4. Moshaverinia, A. Review of the modern dental ceramic restorative materials for esthetic dentistry in the minimally invasive age. *Dent. Clin. N. Am.* **2020**, *64*, 621–631. [CrossRef] [PubMed]
5. Fasbinder, D.J.; Neiva, G.F.; Heys, D.; Heys, R. Clinical evaluation of chairside Computer Assisted Design/Computer Assisted Machining nano-ceramic restorations: Five-year status. *J. Esthet. Restor. Dent.* **2020**, *32*, 193–203. [CrossRef]
6. Jovanović, M.; Živić, M.; Milosavljević, M. A potential application of materials based on a polymer and CAD/CAM composite resins in prosthetic dentistry. *J. Prosthodont. Res.* **2021**, *65*, 137–147. [CrossRef]
7. Awada, A.; Nathanson, D. Mechanical properties of resin-ceramic CAD/CAM restorative materials. *J. Prosthet. Dent.* **2015**, *114*, 587–593. [CrossRef]
8. Beyabanaki, E.; Eftekhar Ashtiani, R.; Feyzi, M.; Zandinejad, A. Evaluation of microshear bond strength of four different CAD-CAM polymer-infiltrated ceramic materials after thermocycling. *J. Prosthodont.* **2022**, *31*, 623–628. [CrossRef]
9. Avram, L.T.; Galațanu, S.-V.; Opriș, C.; Pop, C.; Jivănescu, A. Effect of different etching times with hydrofluoric acid on the bond strength of CAD/CAM ceramic material. *Materials* **2022**, *15*, 7071. [CrossRef]
10. Motevasselian, F.; Amiri, Z.; Chiniforush, N.; Mirzaei, M.; Thompson, V. In vitro evaluation of the effect of different surface treatments of a hybrid ceramic on the microtensile bond strength to a luting resin cement. *J. Lasers Med. Sci.* **2019**, *10*, 297–303. [CrossRef]
11. Reymus, M.; Roos, M.; Eichberger, M.; Edelhoff, D.; Hickel, R.; Stawarczyk, B. Bonding to new CAD/CAM resin composites: Influence of air abrasion and conditioning agents as pretreatment strategy. *Clin. Oral. Investig.* **2019**, *23*, 529–538. [CrossRef]
12. Mine, A.; Kabetani, T.; Kawaguchi-Uemura, A.; Higashi, M.; Tajiri, Y.; Hagino, R.; Imai, D.; Yumitate, M.; Ban, S.; Matsumoto, M.; et al. Effectiveness of current adhesive systems when bonding to CAD/CAM indirect resin materials: A review of 32 publications. *Jpn. Dent. Sci. Rev.* **2019**, *55*, 41–50. [CrossRef]
13. Yoshihara, K.; Nagaoka, N.; Maruo, Y.; Nishigawa, G.; Irie, M.; Yoshida, Y.; Van Meerbeek, B. Sandblasting may damage the surface of composite CAD–CAM blocks. *Dent. Mater.* **2017**, *33*, e124–e135. [CrossRef]
14. Dos Santos, D.M.; da Silva, E.V.; Vechiato-Filho, A.J.; Cesar, P.F.; Rangel, E.C.; da Cruz, N.C.; Goiato, M.C. Aging effect of atmospheric air on lithium disilicate ceramic after nonthermal plasma treatment. *J. Prosthet. Dent.* **2016**, *115*, 780–787. [CrossRef]
15. Tabari, K.; Hosseinpour, S.; Mohammad-Rahimi, H. The impact of plasma treatment of Cercon zirconia ceramics on adhesion to resin composite cements and surface properties. *J. Lasers Med. Sci.* **2017**, *8*, S56–S61. [CrossRef]
16. Liao, Y.; Liu, X.Q.; Chen, L.; Zhou, J.F.; Tan, J.G. Effects of different surface treatments on the zirconia-resin cement bond strength. *J. Peking. Univ. Health Sci.* **2018**, *50*, 53–57.
17. Ye, X.Y.; Liu, M.Y.; Li, J.; Liu, X.Q.; Liao, Y.; Zhan, L.L.; Zhu, X.M.; Li, H.P.; Tan, J.G. Effects of cold atmospheric plasma treatment on resin bonding to high-translucency zirconia ceramics. *Dent. Mater. J.* **2022**, *41*, 896–904. [CrossRef]
18. Vechiato Filho, A.J.; dos Santos, D.M.; Goiato, M.C.; de Medeiros, R.A.; Moreno, A.; da Rocha Bonatto, L.; Rangel, E.C. Surface characterization of lithium disilicate ceramic after nonthermal plasma treatment. *J. Prosthet. Dent.* **2014**, *112*, 1156–1163. [CrossRef]
19. Kim, D.S.; Ahn, J.J.; Kim, G.C.; Jeong, C.M.; Huh, J.B.; Lee, S.H. Influence of non-thermal atmospheric pressure plasma treatment on retentive strength between zirconia crown and titanium implant abutment. *Materials* **2021**, *14*, 2352. [CrossRef]
20. Günal-Abduljalil, B.; Önöral, Ö.; Ongun, S. Micro-shear bond strengths of resin-matrix ceramics subjected to different surface conditioning strategies with or without coupling agent application. *J. Adv. Prosthodont.* **2021**, *13*, 180–190. [CrossRef]
21. Hagino, R.; Mine, A.; Kawaguchi-Uemura, A.; Tajiri-Yamada, Y.; Yumitate, M.; Ban, S.; Miura, J.; Matsumoto, M.; Yatani, H.; Nakatani, H. Adhesion procedures for CAD/CAM indirect resin composite block: A new resin primer versus a conventional silanizing agent. *J. Prosthodont. Res.* **2020**, *64*, 319–325. [CrossRef] [PubMed]
22. Yang, Y.; Zheng, M.; Jia, Y.N.; Li, J.; Li, H.P.; Tan, J.G. Time-dependent reactive oxygen species inhibit Streptococcus mutans growth on zirconia after a helium cold atmospheric plasma treatment. *Mater. Sci. Eng. C Mater. Biol. Appl.* **2021**, *120*, 111633. [CrossRef] [PubMed]
23. Zhu, X.M.; Zhou, J.F.; Guo, H.; Zhang, X.F.; Liu, X.Q.; Li, H.P.; Tan, J.G. Effects of a modified cold atmospheric plasma jet treatment on resin-dentin bonding. *Dent. Mater. J.* **2018**, *37*, 798–804. [CrossRef] [PubMed]
24. Ma, X.R.; Zhu, X.M.; Li, J.; Qi, X.; Li, H.P.; Tan, J.G. Characterization of cold atmospheric plasma-modified dentin collagen. *Dent. Mater. J.* **2022**, *41*, 473–480. [CrossRef] [PubMed]
25. Tekçe, N.; Tuncer, S.; Demirci, M. The effect of sandblasting duration on the bond durability of dual-cure adhesive cement to CAD/CAM resin restoratives. *J. Adv. Prosthodont.* **2018**, *10*, 211–217. [CrossRef]
26. Stancampiano, A.; Forgione, D.; Simoncelli, E.; Laurita, R.; Tonini, R.; Gherardi, M.; Colombo, V. The effect of cold atmospheric plasma (CAP) treatment at the adhesive-root dentin interface. *J. Adhes. Dent.* **2019**, *21*, 229–237.
27. Lata, S.; Chakravorty, S.; Mitra, T.; Pradhan, P.K.; Mohanty, S.; Patel, P.; Jha, E.; Panda, P.K.; Verma, S.K.; Suar, M. Aurora Borealis in dentistry: The applications of cold plasma in biomedicine. *Mater. Today Bio.* **2021**, *13*, 100200. [CrossRef]
28. Henningsen, A.; Smeets, R.; Heuberger, R.; Jung, O.T.; Hanken, H.; Heiland, M.; Cacaci, C.; Precht, C. Changes in surface characteristics of titanium and zirconia after surface treatment with ultraviolet light or non-thermal plasma. *Eur. J. Oral Sci.* **2018**, *126*, 126–134. [CrossRef]
29. Barquete, C.G.; Simão, R.A.; Almeida Fonseca, S.S.; Elias, A.B.; Antunes Guimarães, J.G.; Herrera, E.Z.; Mello, A.; Moreira da Silva, E. Effect of cementation delay on bonding of self-adhesive resin cement to yttria-stabilized tetragonal zirconia polycrystal ceramic treated with nonthermal argon plasma. *J. Prosthet. Dent.* **2021**, *125*, 693.e1–693.e7. [CrossRef]

30. Bormashenko, E.; Chaniel, G.; Grynyov, R. Towards understanding hydrophobic recovery of plasma treated polymers: Storing in high polarity liquids suppresses hydrophobic recovery. *Appl. Surf. Sci.* **2013**, *273*, 549–553. [CrossRef]
31. Zheng, M.; Zhan, L.L.; Liu, Z.Q.; Li, H.P.; Tan, J.G. Effect of different plasma treated zirconia on the adhesive behaviour of human gingival fibroblasts. *J. Peking. Univ. Health Sci.* **2019**, *51*, 315–320.
32. Yoda, N.; Abe, Y.; Suenaga, Y.; Matsudate, Y.; Hoshino, T.; Sugano, T.; Nakamura, K.; Okino, A.; Sasaki, K. Resin cement-zirconia bond strengthening by exposure to low-temperature atmospheric pressure multi-gas plasma. *Materials* **2022**, *15*, 631. [CrossRef]
33. Takamatsu, T.; Uehara, K.; Sasaki, Y.; Miyahara, H.; Matsumura, Y.; Iwasawa, A.; Ito, N.; Azuma, T.; Kohno, M.; Okino, A. Investigation of reactive species using various gas plasmas. *RSC. Adv.* **2014**, *4*, 39901–39905. [CrossRef]
34. Chen, M.; Zhang, Y.; Yao, X.; Li, H.; Yu, Q.; Wang, Y. Effect of a non-thermal, atmospheric-pressure, plasma brush on conversion of model self-etch adhesive formulations compared to conventional photo-polymerization. *Dent. Mater.* **2012**, *28*, 1232–1239. [CrossRef]
35. Papadopoulos, K.; Pahinis, K.; Saltidou, K.; Dionysopoulos, D.; Tsitrou, E. Evaluation of the surface characteristics of dental CAD/CAM materials after different surface treatments. *Materials* **2020**, *13*, 981. [CrossRef]
36. Lise, D.P.; Van Ende, A.; De Munck, J.; Vieira, L.; Baratieri, L.N.; Van Meerbeek, B. Microtensile bond strength of composite cement to novel CAD/CAM materials as a function of surface treatment and aging. *Oper. Dent.* **2017**, *42*, 73–81. [CrossRef]
37. Castro, E.F.; Azevedo, V.L.B.; Nima, G.; Andrade, O.S.; Dias, C.T.D.S.; Giannini, M. Adhesion, mechanical properties, and microstructure of resin-matrix CAD-CAM ceramics. *J. Adhes. Dent.* **2020**, *22*, 421–431.
38. Ahn, J.J.; Kim, D.S.; Bae, E.B.; Kim, G.C.; Jeong, C.M.; Huh, J.B.; Lee, S.H. Effect of non-thermal atmospheric pressure plasma (NTP) and zirconia primer treatment on shear bond strength between Y-TZP and resin cement. *Materials* **2020**, *13*, 3934. [CrossRef]
39. Gale, M.S.; Darvell, B.W. Thermal cycling procedures for laboratory testing of dental restorations. *J. Dent.* **1999**, *27*, 89–99. [CrossRef]

Disclaimer/Publisher's Note: The statements, opinions and data contained in all publications are solely those of the individual author(s) and contributor(s) and not of MDPI and/or the editor(s). MDPI and/or the editor(s) disclaim responsibility for any injury to people or property resulting from any ideas, methods, instructions or products referred to in the content.

MDPI AG
Grosspeteranlage 5
4052 Basel
Switzerland
Tel.: +41 61 683 77 34

Materials Editorial Office
E-mail: materials@mdpi.com
www.mdpi.com/journal/materials

Disclaimer/Publisher's Note: The title and front matter of this reprint are at the discretion of the Guest Editors. The publisher is not responsible for their content or any associated concerns. The statements, opinions and data contained in all individual articles are solely those of the individual Editors and contributors and not of MDPI. MDPI disclaims responsibility for any injury to people or property resulting from any ideas, methods, instructions or products referred to in the content.